practical CLASSICS

& CAR RESTORER

on Classic Auto Electrics

Published by

KELSEY PUBLISHING LIMITED

Printed in Great Britain by
Garnet Dickinson Print Limited
Eastwood Works, Fitzwilliam Road,
Rotherham, S. Yorks S65 1JU
on behalf of Kelsey Publishing Ltd

© 1991

ISBN 1 873098 06 5

Acknowledgements

We would like to thank EMAP National
Publications Ltd for their permission
to reprint these articles. We would
like to thank the main contributors, in
partcular Joss Joselyn, John Williams,
Peter Wallage, David Hill and Peter
Simpson. This compilation was edited
by Gordon Wright of Kelsey Publishing Ltd.

Contents

Introduction

Over the years 'Practical Classics & Car Restorer' magazine has covered a wide range of the aspects of classic car auto-electrics. We have selected what we consider to be the most useful articles and the ones which we believe collectively will give a comprehensive coverage of the subject. Where two articles on the same aspect (eg. valve radio repair) have been included, it has been done so because we have seen that a different emphasis has helped to give a broader coverage.

One of the main purposes for publishing this book is that the highly illustrated nature of 'Practical Classics & Car Restorer' articles makes them so much easier to understand than some of the more academic works on the subject. We hope readers will find them useful, interesting and even entertaining.

GW

All prices quoted in this book were current only at the time that the original articles were written.

Better Distribution

Joss Joselyn delves into overhauling your Lucas distributor.

LUCAS DM2 DISTRIBUTOR

"Early English," is the way an architectural student might describe this archetypal Lucas distributor. Direct ancestor to many modern Lucas units, the DM2 used to be fitted to quite a wide range of cars.

It is a pretty long lived unit, which is a good thing, because a complete overhaul nowadays is a bit difficult as Lucas have stopped supplying a few of the essential bits and pieces!

There is still a lot that you can do, however, so let's start at the beginning and find out just what's possible and what isn't.

For any sort of major overhaul it is best to have the unit out of the car but if you want to avoid re-timing the ignition from scratch first take a few precautions. Take the cap off and note which way round the distributor is located (which side the vacuum unit is positioned) and the exact position of the rotor arm. As a precaution make a simple sketch of these points. Then, provided you don't turn the engine over in the meantime, you can put the unit back later with hardly any disturbance of the timing.

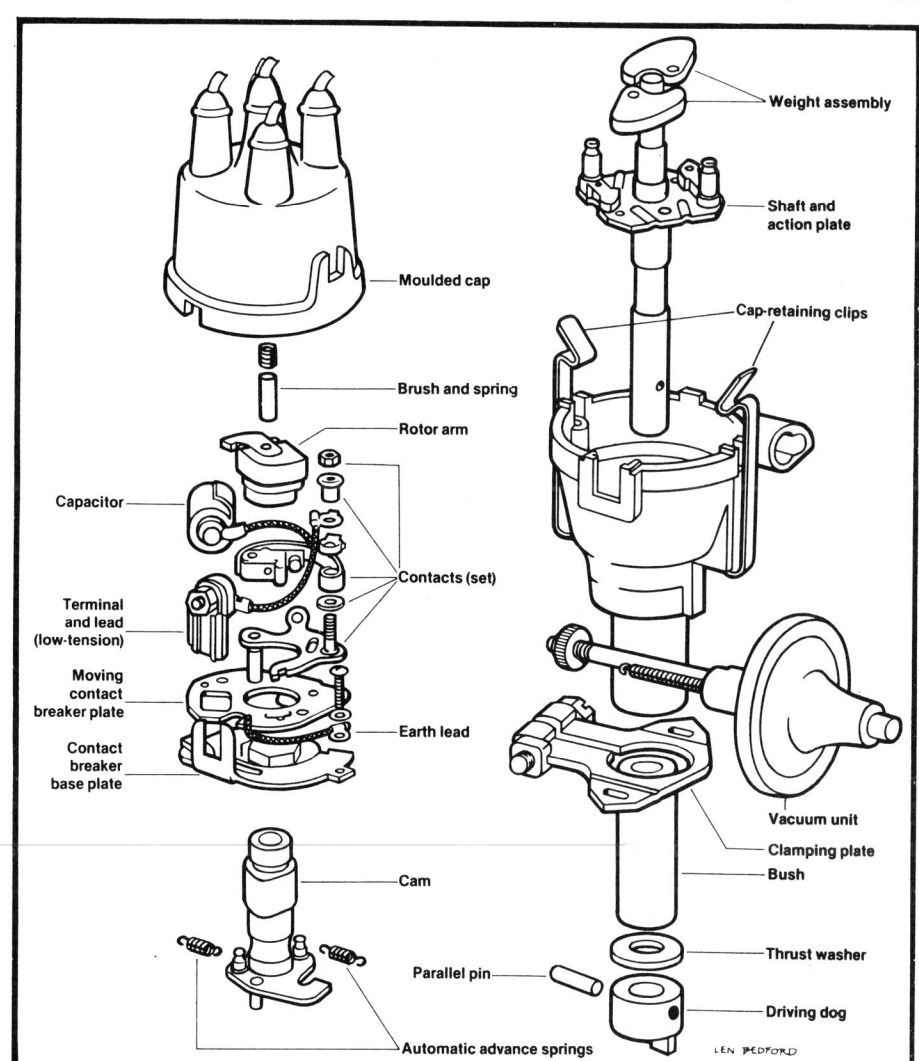

- Moulded cap
- Brush and spring
- Rotor arm
- Capacitor
- Contacts (set)
- Terminal and lead (low-tension)
- Moving contact breaker plate
- Earth lead
- Contact breaker base plate
- Cam
- Parallel pin
- Automatic advance springs
- Weight assembly
- Shaft and action plate
- Cap-retaining clips
- Vacuum unit
- Clamping plate
- Bush
- Thrust washer
- Driving dog

LEN BEDFORD

Tools you will need.

Masking tape • Spanners — various small • Small plain screwdriver • Small Philips screwdriver • Light hammer • Small drift (parallel sided) • Oil can (engine oil or lighter) • Feeler gauges • Narrow nosed pliers • Small flat file • Strobe light — if available.

Parts you will need.

Set of points • Condenser • Others depending upon the condition of the unit overhauled and what can be obtained.

Better Distribution *(Continued)*

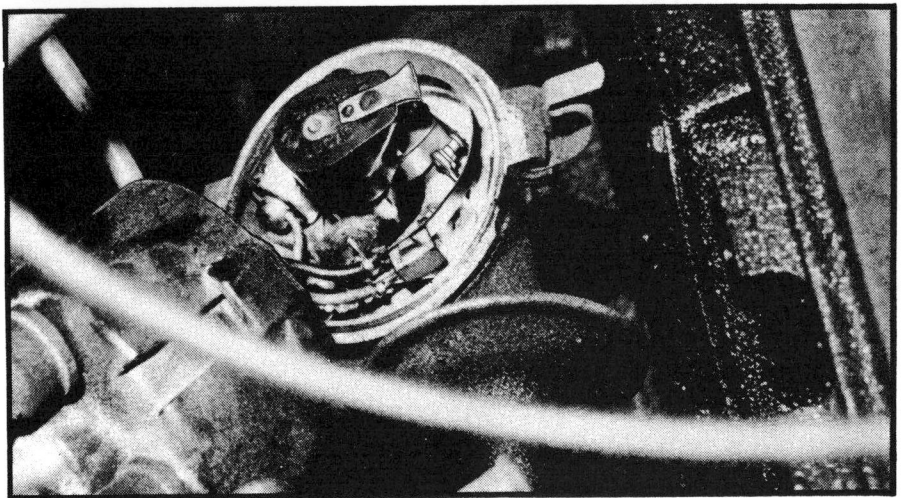

Before removing the distributor from the car, take the cap off and check which way the rotor arm is pointing. Either remember it or make a sketch so it can be put back in the same position.

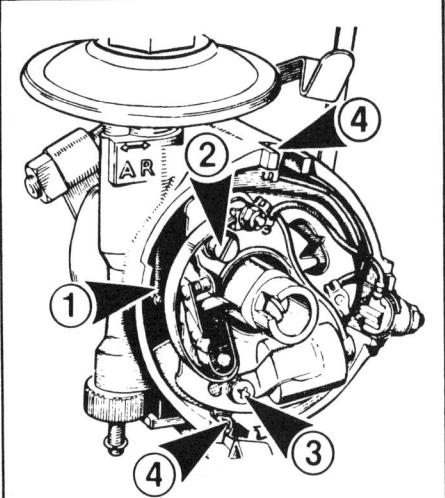

Remove the points, condensor and terminal assembly. (1.) Vacuum advance operated spring. (2.) Points assembly securing screw. (3.) Condenser fixing screw. (4.) Screws securing base plate.

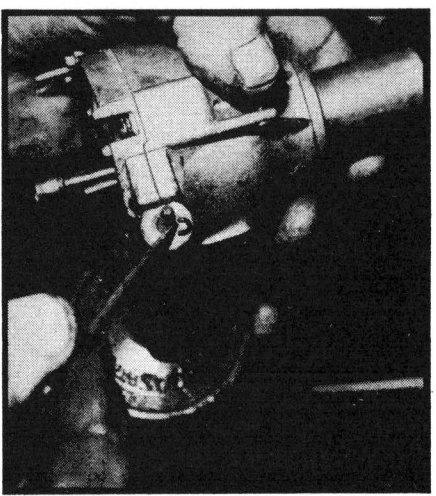

Start removal of the vacuum advance unit by unhooking the spring that links it to the moving contact breaker plate and then use a screwdriver blade to remove the little circlip holding the adjustment nut.

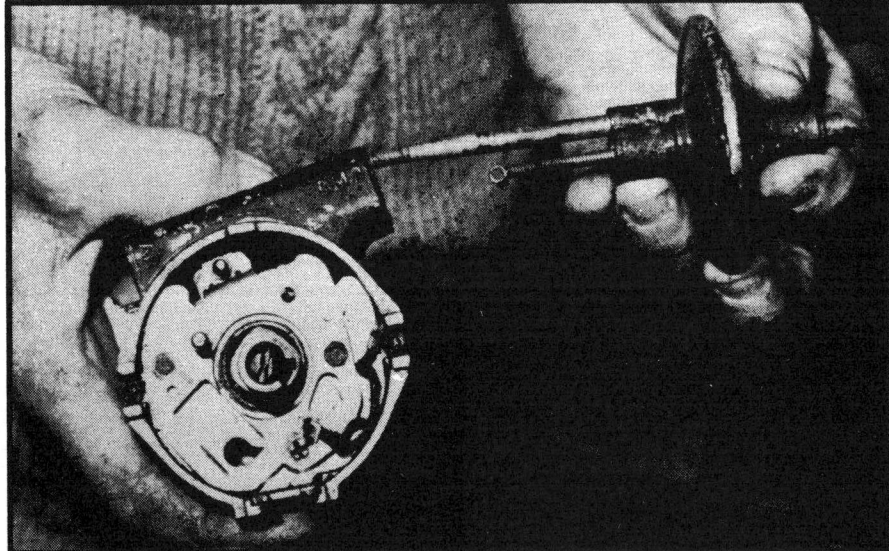

Remove the adjustment nut by unscrewing it, taking great care not to lose the little detent (ratchet) spring which has a nasty habit of jumping. The complete vacuum unit can then be detached.

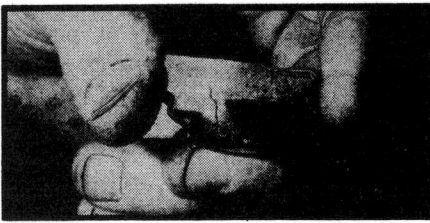

With the unit on the bench, remove cover and rotor arm to inspect for cracks or other signs of damage. The crack on this rotor is typical.

Remove the two screws and lift out the contact breaker and base plate.

If you think you might have trouble getting the plug leads back in the right order, number them before you pull them off the plugs. A piece of masking tape wrapped round and numbered in biro is an easy way to mark the leads, but do not cut notches in them. Apart from the HT leads, the only other dismantling is the vacuum advance pipe from the carburettor or inlet manifold and the LT lead at the side of the unit which is simply disconnected.

At this early stage in the operation it is a good idea to remove the rotor arm and by placing a finger on the central shaft now exposed check the condition of the bushes the shaft runs in. Over a long period these may have worn oval allowing the shaft to float about. If this happens the cam on the shaft will not open the points to the correct gap when the engine is running no matter how carefully they have been set.

This irregular gap at the points will not help smooth running, performance or starting. As the wear gets worse a tendency for the engine to mis-fire will become increasingly apparent.

If there is any appreciable side-play in the shaft it is a good idea to obtain a replacement unit and forget any plans to overhaul the original.

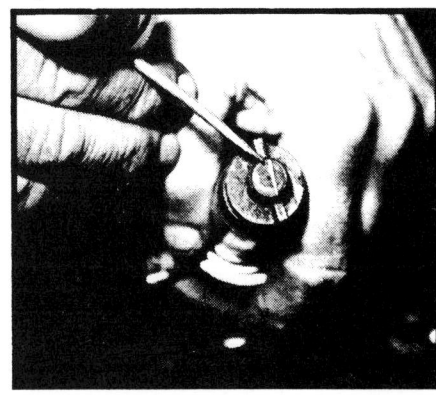

Before removing the drive dog at the bottom of the spindle, mark its position in relation to the shaft.

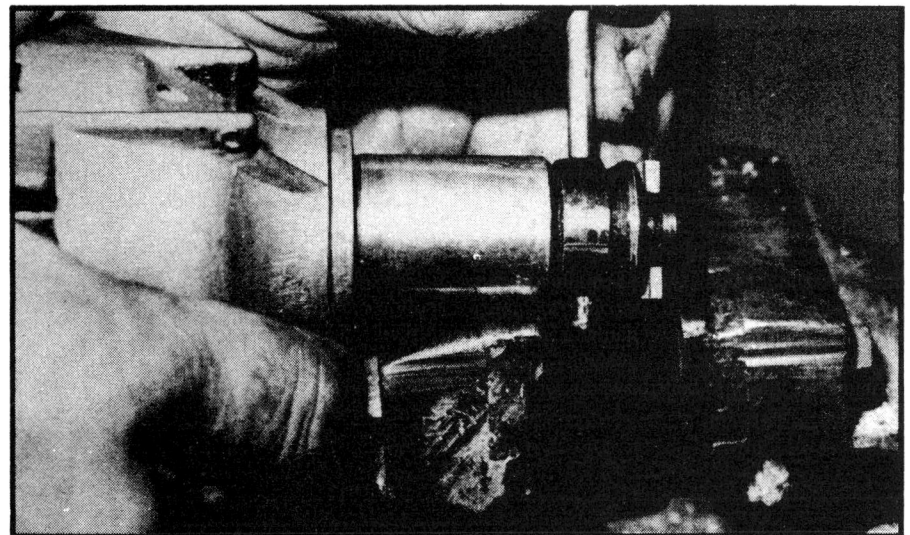

Use a hammer with a thin parallel sided drift to tap out the securing pin. No need to drive it right out; just enough to clear the shaft.

REMOVAL

If all is well with the bushes and the shaft the distributor can now be removed. At the point where the distributor emerges from the crankcase or cylinder head there will either be a clamp or a flange. Before slackening the studs securing the unit to the head or crankcase try to make a mark which will indicate its position in relation to the engine. The distributor can then be lifted out when the studs have been removed.

The reason for making sketches and leaving marks to fix the position of the rotor arm in relation to the distributor body and the distributor in relation to the engine is to provide a starting point for establishing the ignition timing when you are ready to re-install. Working on the principle that if the engine ran with such settings before the distributor was disturbed it should run when all the variables are returned to the starting point — that is, if luck is on your side.

On the bench, check first the cap and the rotor for damage or signs of tracking. For this they'll need to be cleaned up thoroughly with petrol and then inspected very closely. Hairline cracks are not easy to see although tracking may well stand out as a frost-like etching in a line and travelling to earth. Look also at the contacts inside the cap for evidence of burning and also the brass segment on the rotor for the same thing. Renew either unit if it is at all suspect. Look inside the 'towers' on the outside of the cap to see if there are signs of burning or corrosion. If there is, try cleaning it out but fit a new cap if in doubt. Check the little spring-loaded carbon brush in the centre. It should be free to move and not binding and still springy in its action.

Our series of photographs and captions describes the dismantling process which is really very simple. Take heed of the warning about the little detent spring which gives the ratchet effect to the advance-retard micro-

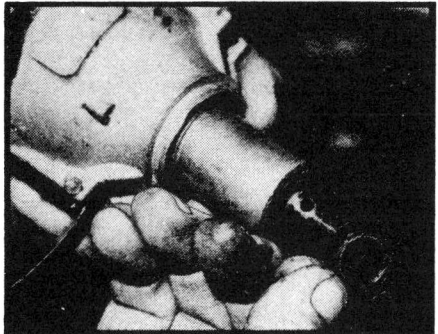

Be careful when you remove the dog to separate and not to lose the bronze or fibre thrust washer.

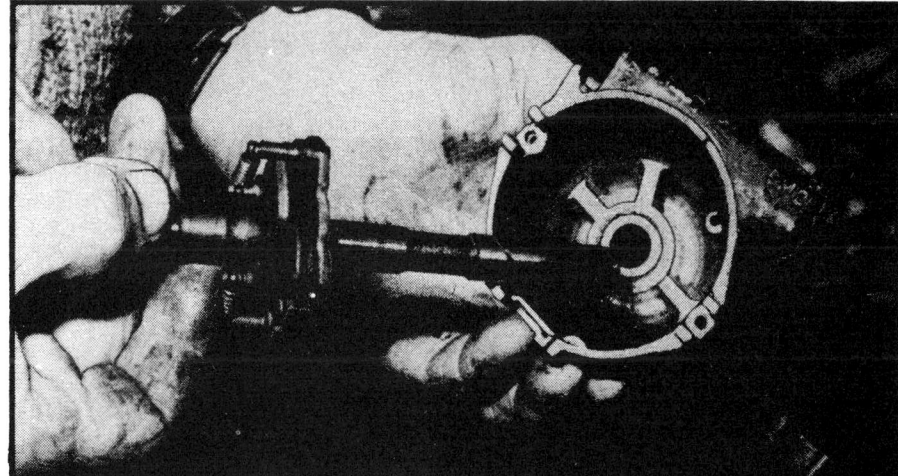

Carefully file off any burrs around the pin hole in the shaft and it should then pull out from the distributor body.

adjuster. It may leap into the darkest corner of the workshop if the adjuster nut is removed too exuberantly. The point about marking the drive dog before removal is that putting the offset D-drive back the wrong way will upset the ignition timing when you refit the distributor. If there is a skew gear drive instead of the D-drive, and there is on some cars, in theory it shouldn't make any difference which way round it is fitted. In practice, however, the hole for the securing pin is not always drilled centrally — so mark it anyway.

There are no hard and fast rules about this. The cap and rotor must be changed if they are at all defective as already described. It is worth fitting a new set of points and another condenser in any case. The older points which were in two parts and had that confusing little assembly of parts on the post at the fixed end of the points spring, are no longer supplied by Lucas and you'll have to fit the later one-piece assembly in any case.

If you can find anything wrong with the breaker and base plate, you can renew this; the parts are still available. You will be looking for wear which will be denoted by a sloppy fit between breaker plate and base or bearer plate or perhaps a worn or loose pivot post.

If there is any problem with the drive dog — whether it is offset D or skew gear, you can get a replacement, but it is at this point that you may run into difficulties. Spares are no longer carried for the centrifugal advance assembly and if you want a new cam, a new shaft, centrifugal weights or springs, a new bearing bush or vacuum unit, you will probably have to get a new distributor complete. It is just possible, however, if you look around, that you may find someone who still has these parts in stock. It may be worth the search if you need only something small, like new centrifugal advance springs but generally, if you find there is excessive wear in the spindle or the major parts need renewal, a new distributor might be the better solution.

Before you dismantle the vacuum advance mechanism, note the position of the springs and mark the pin which is adjacent to the offset stop and the stop itself, as shown.

Better Distribution

(Continued)

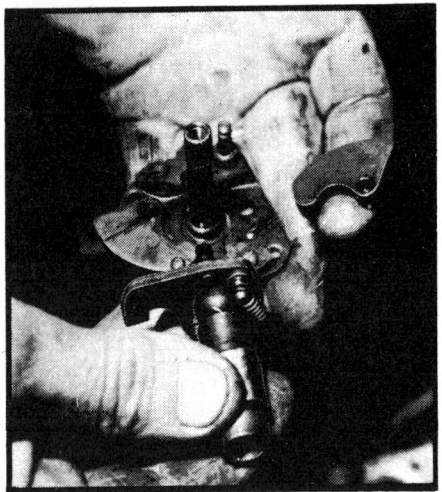

Dismantle by removing the locking screw from the centre of the rotor arm drive, carefully unhooking the centrifugal weight springs and lifting off the cam spindle.

REASSEMBLY

This is basically the reverse of the dismantling process but there are a few points which should be noted.

First, if you forgot to note the position of the springs and the offset stop in the centrifugal advance mechanism, all is not lost. Looking down on the assembly from the top and with the rotor positioned at six o'clock, the offset stop should be on the right.

Assemble the weights, springs and cam onto the action plate on the shaft and lubricate it all sparingly with engine oil. You will, of course, already have cleaned up all the parts and all that remains now is to check that they operate smoothly.

Oil the shaft before reinstalling it and make sure that the thrust washer is replaced and that

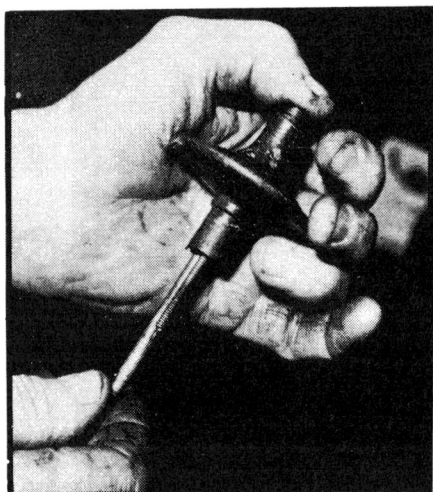

This is the method of checking the action of the vacuum advance by pulling on the diaphragm operating arm. A thumb is then used to seal the end. When the thumb is moved, the diaphragm should reassert itself.

DM2 usage and availability.

Lucas have informed us that the DM2P4 Distributor was introduced in April, 1952, and was adopted as original equipment on a wide range of makes and models. It was replaced by the 23D/25D range of distributors at the beginning of the sixties. Because of the diversity of usage and the time lapse Lucas say that it is now impractical to compile a comprehensive list of users, but stress that while some spares are in short supply they are willing to help with individual enquiries. Try your local Lucas stockist first, they may be able to supply all the parts required for an overhaul of the assembly, but if problems arise write to:- Lucas Electrical Ltd., Parts and Service Division, Great Hampton Street, Birmingham B18 6AU or the one make club for your car.

The Lucas type number is generally in raised letters on the main casting.

---

the dog is fitted the right way round by noting the match marks.

Check the movement of the contact breaker base plate on the bearer plate and lightly oil it before reinstalling.

When reassembling the vacuum mechanism, don't forget to refit the little detent spring behind the knurled nut of the micrometer adjustment. Fit the adjuster half way along its thread and refit the little circlip, using a pair of needle-nose pliers to close it up again tightly.

Assemble all the components above the contact breaker base plate, making sure that the various connections are all clean and that the insulation on the wires is sound. If it is damaged or chafed, fit a new one. Clean the faces of the new points before fitting to get rid of preservative that is sometimes applied and which can inhibit their action.

The final reassembly job is setting the contact breaker gap. First the contact breaker plate securing screw is slackened, the appropriate feeler gauge (0.015in. in this case) is inserted between the points and the gap adjusted by inserting and turning a screwdriver blade in the adjoining slot. When the feeler is a tight sliding fit, re-tighten the securing screw.

If the old rotor arm is being refitted, give the contact edge a clean by rubbing it on the side of a tyre. Finally apply a thin smear of grease to the cam and a drop of oil to the pivot of the moving contact. Normally another couple of drops would go between cam and base plate and down the centre of the top of the cam under the rotor but these have already been attended to when reassembling.

Adjust the points gap accurately to 0.015in. as described in the final caption.

Refit the distributor, turning it to get the position noted before removal and rotating the rotor to the correct angle. Provided the engine has not been turned since removal, the timing should be close to what it was. To be on the safe side, if you can, check the timing finally with a strobe light. □

Delco distributors are fitted to most Vauxhalls, but not only to them. Although made by a General Motors company, they have also been used on Triumph, Ford and Chrysler, as well as on Bedford and Opel. There are two main ranges — the D200 and the D300, the principal difference between the two being that the D200 range has the centrifugal advance and retard fitted underneath the contact breaker, while on the D300 it's just underneath the rotor, on top of the contact breaker plate.

The example we chose to overhaul came out of a MkIII Triumph Spitfire, one of the D200 type, but fitted with a later type of contact breaker. This can be stripped down and overhauled but it is worthwhile first finding out just what replacement parts are available in your area. All the fast moving bits – cap, rotor, contact breaker set and condenser – are obtainable almost anywhere, but bob weight springs, shaft, bearings, etc., could be a very different story.

It is best to make an initial check by dismantling the top end of the unit and trying the shaft for lateral movement. If there is a lot of play, it means it's likely that the whole of the unit is worn and a factory replacement is the best remedy. The probability is that you

This is the complete lump as it came out of the car. Note the tachometer drive immediately under the distributor body.

Doing up a Delco

Joss Joselyn starts our new engine ancillaries overhaul series with the Delco D200 distributor.

won't be able to buy a new shaft/advance and retard unit, bearings, etc., but if you can, it's still worthwhile working out the cost of this and all the other parts and comparing it against a factory unit. With this sort of wear, an extensive home overhaul probably won't be worth while. If spindle and bearings seem sound, the more limited overhaul procedure shown in our photographs is the best answer.

Dismantling

Working on the well known principle that any mug can take something apart, and that reassembly is the tricky part, we have only shown the latter in any detail. If you need more visual detail of dismantling, you could always follow the photographs in reverse.

After removing the cap, the rotor simply pulls off. The contact breaker comes out next; simply take out the locking screw, lift the assembly off its pivot pin and disconnect from the terminal assembly. Another single screw will then release the condenser.

Take off the vacuum advance unit next. It is held by two screws, but check just which screws hold what before taking them out. Three screws altogether hold the vacuum unit, the contact breaker base plate, the two pivoting cap clips and the earthing lead. The exact arrangement varies a little from model to model, but the general principle of dismantling is the same. Do not forget to unhook the vacuum control connecting arm from the contact breaker baseplate before

withdrawing it through its slot in the housing.

Normally, the next step would be to remove the tachometer drive gear (where this is fitted); but, because on this unit it was not proposed to change the main shaft (there was nothing wrong with it or the bearings), we omitted this step. If you do want to take it off, you have to lever out the staked plug from the end of the housing using a small screwdriver.

There are no real difficulties in dismantling, but levering the points away does help release the two connections.

The gear can then be extracted, taking care not to lose the small thrust washer.

The final dismantling stage — should you want to do it — is to remove the mainshaft assembly, including the centrifugal advance and retard bobweights. It's done by driving out the pin fitted through the gear or coupling at the base of the shaft. If you plan to reassemble the coupling to the shaft, mark them before dismantling.

Knock the pin out carefully using a thin 'parallel' punch and check the shaft carefully for any burrs. If there are any, remove them before pulling the shaft up through the bearings. Retain the drive components because, should you be able to find a new shaft anywhere, it comes without a driving gear or coupling.

In these circumstances, the hole drilled for the pin in the old coupling can be used as a guide to drill the new shaft. The procedure is to fit the upper thrust washer first, and then slot the shaft into the housing. Put on the lower thrust washer and the coupling or gear, and drill the hole so that the shaft has an end

Doing up a Delco/continued

float (measured between washer and housing) of between 0.002in. and 0.005in.

If you fit a new gear or coupling as well, you've got a problem. Drilling exactly at right angles through both coupling and shaft, centrally, and at the same time ensuring the correct end float, is not easy. Most people would settle for an exchange unit.

In theory it is also possible to fit new bearings — driving out the old ones and pressing in the new — theoretically, because you're unlikely to get the bearings. If it comes to fitting bushes and shaft, even if you could get them, it would almost certainly be cheaper to get a factory replacement.

Reassembly

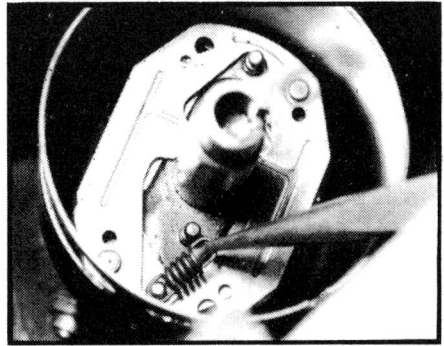

You can clean up the bob weight mechanism without actually removing it, using cleaning fluid and a paintbrush. Here the bobweight springs are re-located — an operation which requires some care.

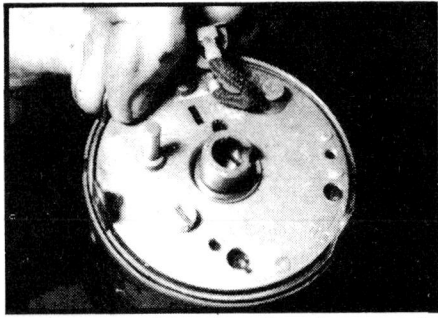

You cannot position the points plate wrongly because the three fixing lugs are unevenly spaced.

After washing off all the bits in a bath of cleaning fluid, start putting it back together by inserting the bob weight springs. Getting these in and out is done with a pair of needle-nose pliers, but it isn't easy. If you're not very careful the spring will zap away across the workshop and be lost for ever. If you can get new ones, it's a good idea to fit them. Once they are installed, oil the mechanism and ensure it works smoothly.

If the tachometer drive gear has been removed, this goes back next. Officially, it should be liberally coated with 'Alvania' No. 2 grease or an equivalent; good quality high melting point grease should do the trick. A new plug is then press-fitted into the shoulder inside the housing and staked in six places.

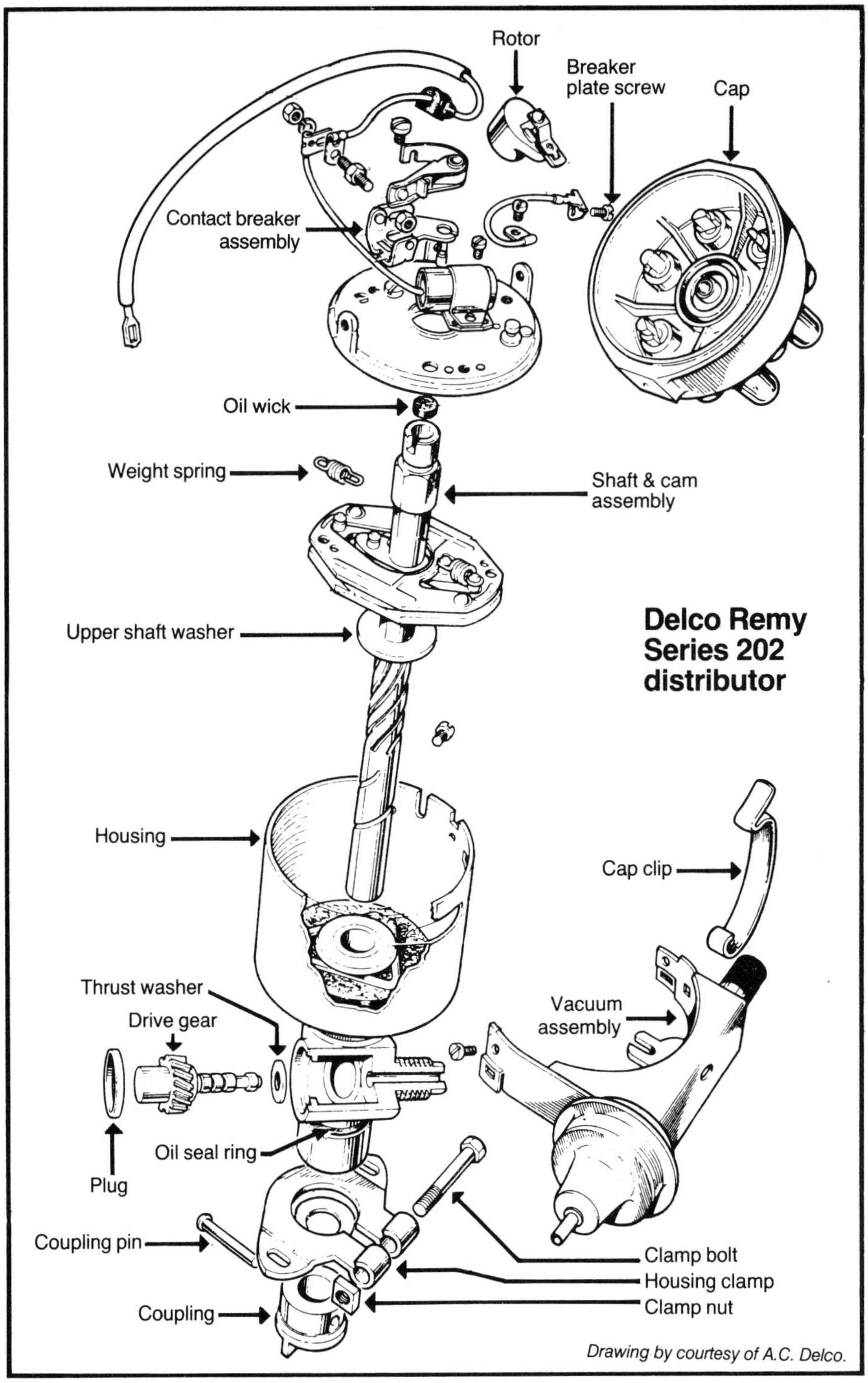

Delco Remy Series 202 distributor

Rotor

Breaker plate screw

Cap

Contact breaker assembly

Oil wick

Weight spring

Shaft & cam assembly

Upper shaft washer

Housing

Cap clip

Thrust washer

Drive gear

Vacuum assembly

Oil seal ring

Plug

Coupling pin

Clamp bolt

Housing clamp

Clamp nut

Coupling

Drawing by courtesy of A.C. Delco.

Tackle the contact breaker plate next. It can be checked quite easily to see that the plate is free to move but not 'wobbly.' Similarly a check should be made of the vacuum advance unit before refitting it. Depress the operating lever and block the inlet firmly with your thumb. If the operating lever stays put until the thumb is removed, it's working, and there should be a partial vacuum which you can hear when you take your thumb off. If either contact breaker plate or vacuum unit is defective, you'll need a replacement. Breakers yard bits are a possibility, of course,

but once again you'll probably have to consider a replacement distributor.

Refitting these two components is not generally difficult and you can't get it wrong because the fixing holes in the body are unequally spaced. Don't forget that two of them also hold the cover clips and one takes the earth lead. One screw holds only the contact breaker baseplate.

Fit the condenser next — a matter of one screw and two little 'pips' to locate it properly. This is followed by the new contact breaker set and, although earlier distributors

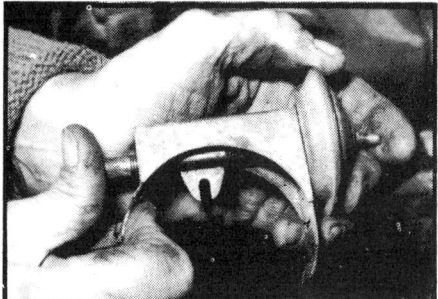

The action of the vacuum advance can be checked with it off the car (see text).

The mounting holes of the contact breaker base plate and vacuum unit all match up, and include the earth lead and two cap clips. Note how the vacuum operating arm connects to the contact breaker base plate.

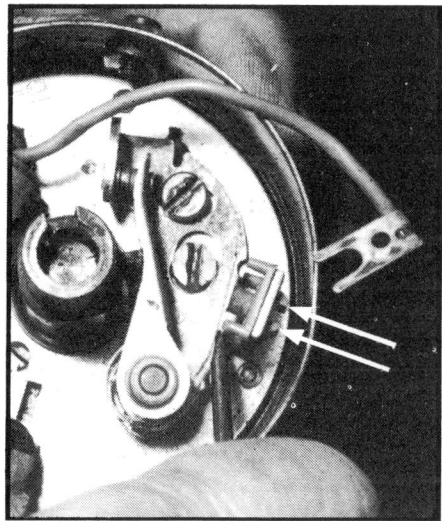

Here the plastic insulator has been fitted, the two ears pushed through the window (arrowed) in the fixed contact, and the condenser lead is poised ready to slot in. The points spring blade is levered away to allow the condenser lead and low tension lead to be slotted in to complete the terminal assembly.

Points adjustment is carried out in the normal way.

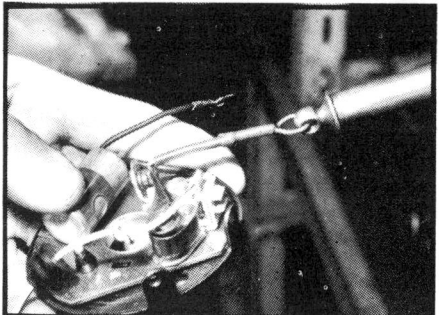

This is the recommended method of measuring contact point pressure using a spring balance.

Lubrication is important and here oil from a can is dribbled onto the wick in the centre spindle and a smear of grease goes onto the cam.

originally had a tedious arrangement of nuts, washers, etc., the later type shown in the drawing is the modern substitute.

First the two ears on the insulator (A) are pushed through the window of contact point (B), pinching them with the fingers if necessary. Install the fixed point (B) onto the pivot pin and fit the locking screw (C). Fit the terminals of the condenser lead (D) and the LT lead (E) onto the centre pin of (A). Fit the moving contact (F) edging it down over the pin and locating the spring carefully between the little pegs on the insulator.

The points gap is set in the usual way with the rubbing block on top of the cam, the gap measured with feeler gauges and adjustment carried out after slackening the locking screw. On the older types the fixed point is adjusted by levering a screwdriver blade in the slot provided; on later series, adjustment is carried out conveniently using the adjuster screw provided.

Delco recommend an extra check, this time of the pressure on the contact breaker points. It's important they say, because if it is too strong, there will be excessive wear of the rubbing block, cam and contact points. If it is too weak, high speed points bounce will result and this, in turn, will cause arcing and burning at the points and a consequent misfire.

A spring balance is used for the check, hooked over the moving contact arm as near the contact as possible. It is pulled at an angle of 90 deg. to the points surface and, just as the points separate, the scale reading should be between 17 and 21oz. If the pressure is excessive, carefully pinching the spring will reduce it. If it is insufficient, it can be increased by taking the points out of the distributor and bending the spring away from the arm. Even new points can have the wrong spring pressure.

Before refitting rotor and cap, lubrication can be carried out. This consists of lightly greasing the cam surfaces with petroleum jelly. Add a few drops of light engine oil to the felt wick in the top of the breaker cam. On the 200 Series there's a small hole through which a few drops of oil should go, while all types have a ¼ in. hole through which about a teaspoonful of thin oil should be trickled. Finally, a few drops should go into the fibre bush in the contact breaker arm pivot.

Rotor and cap

These two items both need cleaning and checking before refitting. Clean off any thick and encrusted oily dirt from the outside of the cap with petrol, polishing off finally with a clean dry rag inside and out. Look for any signs of damage — chips or cracks — and for any traces of 'tracking'. This appears as an etched line which follows the path of a scratch or pattern of dirt, allowing leakage of high tension current to earth. If any is found it means changing the part (cap or rotor).

If the aluminium contacts inside the cap need cleaning, use a petrol soaked rag; don't scrape them or use abrasives as this will increase the gap. If these contacts, or the centre contact are worn, fit a new cap.

Clean dirt off the edge of the rotor segment by rubbing it on the side of a tyre; don't scrape it or use emery cloth. Check also the height of the spring contact. From the top face of the rotor to the top of the domed contact should be 13/32in. (1.0 cm.)

Check the HT cable sockets in the 'towers' on the cap. Bad burning from a past faulty contact means fitting a new cap. Check also that the sleeves on the cable ends have the fingers to engage the grooves in the sockets. Clean up with cleaning fluid before refitting.☐

The Generation Game

Joss Joselyn explains the overhaul of a Lucas Dynamo

LUCAS C40 DYNAMO

The Lucas dynamo is a simple and robust unit and not a lot goes wrong with it. There are two ways in which potential problems may come to your notice, however. First, you may spot that evil red eye winking at you from the dash panel or second, a sagging battery may indicate trouble in the charging circuit.

Once you have confirmed that it is the dynamo that is at fault, an investigation and, if necessary, an overhaul is well within the capabilities of the home mechanic. Apart from cleaning, there are really only three things you can do. You can fit new brushes, you can clean up the commutator and you can fit new bearings.

Some of the major points can be seen in our photographs and all the major components can be identified from the diagram. Removing the unit is no problem; it is just a matter of gently pulling off the two electrical connections at the back, unbolting the two mountings and the adjustment strap, separating the fanbelt from the pulley and lifting the unit off.

DISMANTLING

The job is best done with the body or yoke mounted in the vice and all that really holds the dynamo end brackets and body together is a pair of long through bolts. Before extracting these, put alignment marks on both end brackets and the body to ensure correct reassembly. It is possible however, that the end brackets may both have a pip cast in and this corresponds with small recesses cut in the rim of the body. In this case, marking is unnecessary. If the end brackets (which are aluminium castings and should be treated gently) stick in position, tapping them with a hide mallet will enable them to be removed.

With the dynamo yoke clamped firmly in the vice, dismantling is a matter of extracting two through bolts and pulling off the rear end bracket which carries the brushes.

Tools You Will Need

Vice • Two screwdrivers • Hide hammer • Fine sandpaper • Ammeter with low reading facility • Suitable spanners for the mountings and the nut securing the pulley • Hammer • Slim punch • Fine, slim file.

The Generation Game

COMMUTATOR

Look first at the condition of the commutator. If this is very badly worn, either by deep scoring or pitting usually caused by prolonged use with worn brushes, no repair is possible and you will need an exchange unit. The only other alternative is to obtain another unit from a breakers yard but unless you can dismantle this before you buy, you could end up with a commutator just as bad as yours.

To inspect the commutator properly, rub all the dirt off using petrol or cleaning fluid. Normally this is all that is necessary. Light scoring and wear can be taken off with a fine strip of sandpaper, using this by wrapping it around and pulling the ends to and fro. Do not be tempted to use emery cloth instead.

If the copper of the commutator is quite thick and the scoring is too severe to be removed by this method, your local auto-electrician will be able to turn the commutator in a lathe and clean it up by taking off a very light skim of metal.

The commutator can be inspected for wear and if it is as badly worn as this one, it is beyond repair.

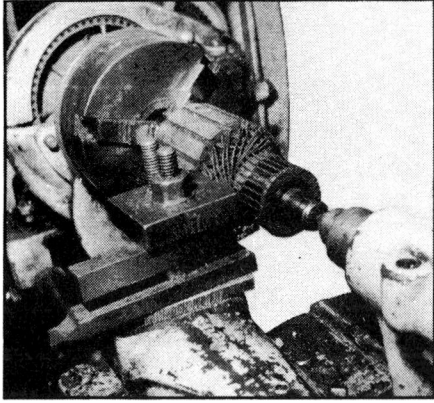

If the commutator segments are considerably worn, the only way it can be restored is by having it skimmed on a lathe. The practicality of this depends on the thickness of metal and your local auto-electrician is the man who can advise.

After either cleaning up with sandpaper or having the commutator skimmed, the depth of the undercutting between the segments should be checked. The slots should be 1/32in. deep and, if skimming the metal has made them less than this, they will have to be re-cut. For this an old hacksaw blade can be used. Reduce it to the correct thickness by grinding it and then carefully cut out the mica between the segments until there is a good clean square-cut groove in every case.

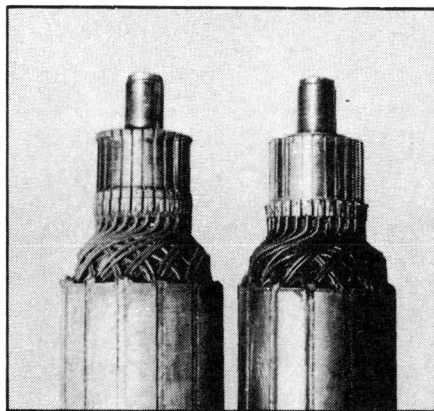

Here, for comparison, are a very badly worn commutator and a restored one.

BODY AND FIELD WINDINGS

Use a bath of cleaning fluid and an old paint brush to clean the body and field windings up inside and out. Then, when all the carbon dust and dirt from the brushes has been removed, check the condition of the insulating wrappings. If they are damaged, there is not much you can do about it and an exchange unit is the best bet.

The only way you can check the windings is with a low-reading ammeter. Connect the ammeter and the field coils in series with a 12V car battery. You should get a reading of around 2 amps. If you get a higher reading, it means there is a partial short and the only solution is an exchange dynamo.

Look at the little wire between the windings. This must not be broken or have come adrift and it should have a small piece of insulation underneath it to prevent the through bolt touching.

Clean up the inside of the body and check the windings for damage to the insulative wrappings. Damage to these or if the link wire shown has broken or come adrift, will mean a new unit.

FRONT END BEARING

Check the condition of the ball race bearing by waggling the shaft in the end plate to detect lateral play. Only if this is detected, is it essential to remove the pulley and front end bracket from the armature and replace the bearing.

The nut securing the pulley comes off first and this is best done holding the armature in the vice. The pulley can then be levered off using a pair of screwdrivers on opposite sides of the shaft, inboard of the pulley. Carefully remove the Woodruff key from its slot in the armature shaft and then support the back of the end bracket against the open jaws of the vice and tap the armature shaft through and out using a hide hammer.

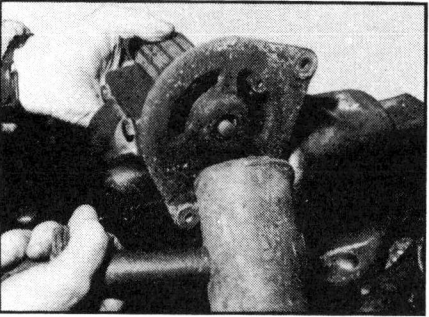

After removing the pulley and extracting the little Woodruff key, the armature can be held between the jaws of the vice to support the end plate while the spindle is knocked through with a hide hammer to avoid damaging the threads.

Turn the end bracket over and it will probably be found that the bearing assembly is held in place by an internal circlip. Prising this out with a screwdriver will enable the bearing assembly to be removed. Note the order of all the retaining plates, washers etc. and which way round they are fitted. Install the new bearing and depending on its condition possibly a new felt washer as well.

Earlier dynamos had a different type of bearing retainer. This was a plate fixed at three points by rivets. The method of extraction here is to drive out the old rivets using a hammer and punch which will enable the plate to be removed, followed by the bearing, a corrugated washer and a felt washer.

On many Lucas dynamos the ball race is locked in by means of an internal circlip which can be prised out.

This releases all the retaining plates and washers as well as the bearing and these must be taken out carefully, noting the sequence of assembly as well as which way round each component goes.

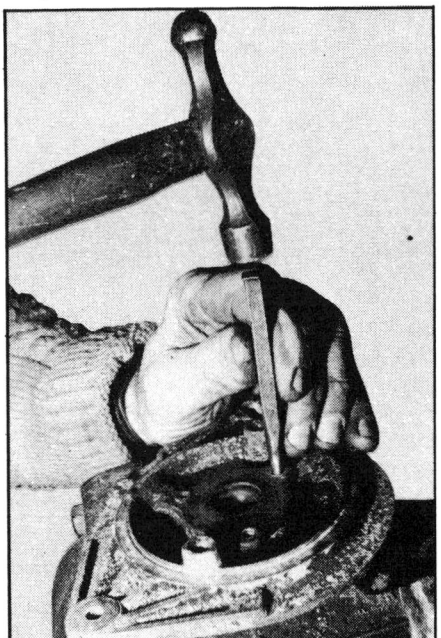

This is an alternative (earlier) method of bearing retention. The retaining plate is riveted and a suitable punch is used to drive them out.

Here are the component parts found under the retaining plate — bearing, corrugated washer and felt washer. Use new rivets when reassembling.

The bearing will be changed, of course, packing it with grease before fitting. It may also be a good idea to fit a new felt washer. If this is difficult to obtain, the old one can be re-constituted by washing in petrol and squeezing gently until it is soft and fluffy again.

New rivets will be needed on reassembly and these can be obtained along with the bearing and felt washer from a Lucas main agent. The rivets are supplied specially for the job and they secure the plate when their open ends are spread with a blunt punch or better still a 'turnover' punch.

REAR END BEARING

The bronze bush rear bearing is checked for wear in the same way as the front one — by testing for play. If it has to be changed, the old one is probably easiest removed by the brutal method of splitting it. This is simply a matter of chopping into the edge of it with a small cold chisel (or an old screwdriver if you must) but do remember that the end plate is an aluminium casting and should be treated with care.

To extract the old bronze bush bearing from the rear end bracket, cut through the wall using a small sharp chisel.

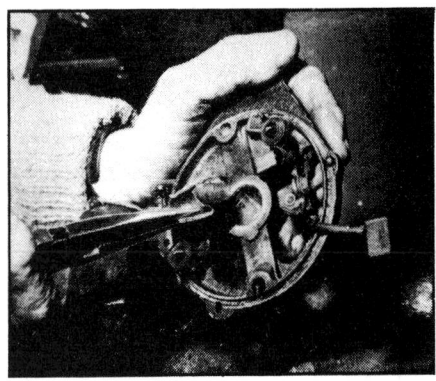

Once the wall has been cut through, the old bush can be pulled out.

Use the vice as a press to fit the new one with a suitable piece of tubing employed as a mandrel.

The Generation Game

A more scientific method of removal is to screw in a 5/8in. tap and when this has secured a strong grip, pull out the bush.

Clean out the housing before fitting a new bush. When this is obtained, it must be soaked in engine oil for 24 hours before fitting although it can, as an alternative, be soaked for a couple of hours in very hot oil. The best method of fitting the bush is to use the vice as a press and a suitable piece of tubing or rod will be necessary as a mandrel. A socket with a suitable size outside diameter is another alternative to use as a mandrel. Take great care to push the bush in straight or it will distort.

BRUSHES

Wash the rear end bracket in petrol or cleaning fluid, just as all the other parts were cleaned and then fish the brushes out of their holders after lifting the tail of the springs clear. A worn brush can be seen compared with a new one in the photographs and the critical dimension that Lucas mention is 9/32in. If the brush length is less than this it should be changed. We would go further and fit new brushes if the old ones are worn to less than 3/8in. or even 1/2in., worn brushes can permit arcing which will pit the commutator and may also stick which will stop the dynamo charging.

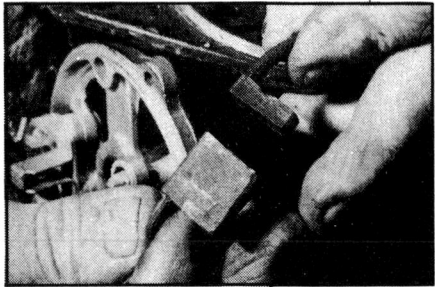

Once the rear end plate has been cleaned up, the brushes can be inspected. Here a new brush is compared with a badly worn one. Fit new brushes if they measure less than 3/8in.

This is the method of reassembling the new brushes to hold them clear of the commutator when re-building. Note spring tip bearing on side of the brush.

CONTINUED ON PAGE 25

INSURANCE BROKERS REGISTRATION COUNCIL
REGULATED FOR INVESTMENT BUSINESS
INDEPENDENT FINANCIAL ADVISER
CJH
C.J.H. GROUP
Incorporating Classic Car Motor Polices
FOR
ALL YOUR INSURANCE REQUIREMENTS
• FIRE • ACCIDENT • MOTOR
• MARINE • LIABILITIES
• HOUSEHOLD • COMMERCIAL
• CLASSIC VEHICLES • CHERISHED COLLECTIONS • WEDDING HIRE
• MOTOR TRADERS • LIFE & PENSION
• MORTGAGES • FINANCE
• INVESTMENTS • ALL INSURANCE ARRANGED AT COMPETITIVE RATES
CLUB MEMBERS DISCOUNT AVAILABLE
PRIVATE HEALTH CARE SCHEME AT DISCOUNT RATES TO CLUB MEMBERS
SPECIAL HOME CONTENTS INS SCHEME
Classic House, 365a Limpfield Road, Warlingham, Surrey, CR6 9HA
Tel: 0883 627491/4 Fax: 0883 627495

SO MUCH FROM SO LITTLE
BOYER BRANSDEN ELECTRONIC IGNITION
FREE IGNITION OFFER PLEASE SEND FOR DETAILS
Here's proof that good things come in little packages: The High Energy Electronic Ignition from Boyer Bransden! BIG performance. BIG m.p.g. figures. BIG value all round! All from an ignition that's smaller than a pack of cigarettes.
It's easy to fit to almost any make of car. In minutes you'll get better starting, smoother running and less tune-up.
What else would give you so much from so little? And at only £18.98 inc. VAT and post the price is the smallest you'll find anywhere!
LIFE TIME GUARANTEE
Access
VISA
For leaflet please send SAE.
Boyer Bransden Electronics Ltd.,
Dept.PC3 Frindsbury House, Cox Lane, Detling, Maidstone, Kent ME14 3HE, Tel:(0622) 30939

KEEPS ITS SPARK
STRONGER FOR LONNN NNNNNNNNGER
CHAMPION
DOUBLE COPPER
FASTER COLD STARTING
BETTER ACCELERATION RESPONSE
FOR UNLEADED OR LEADED FUEL
NO GAP ADJUSTMENT NEEDED
LOWER EMISSIONS
LONGER PLUG LIFE

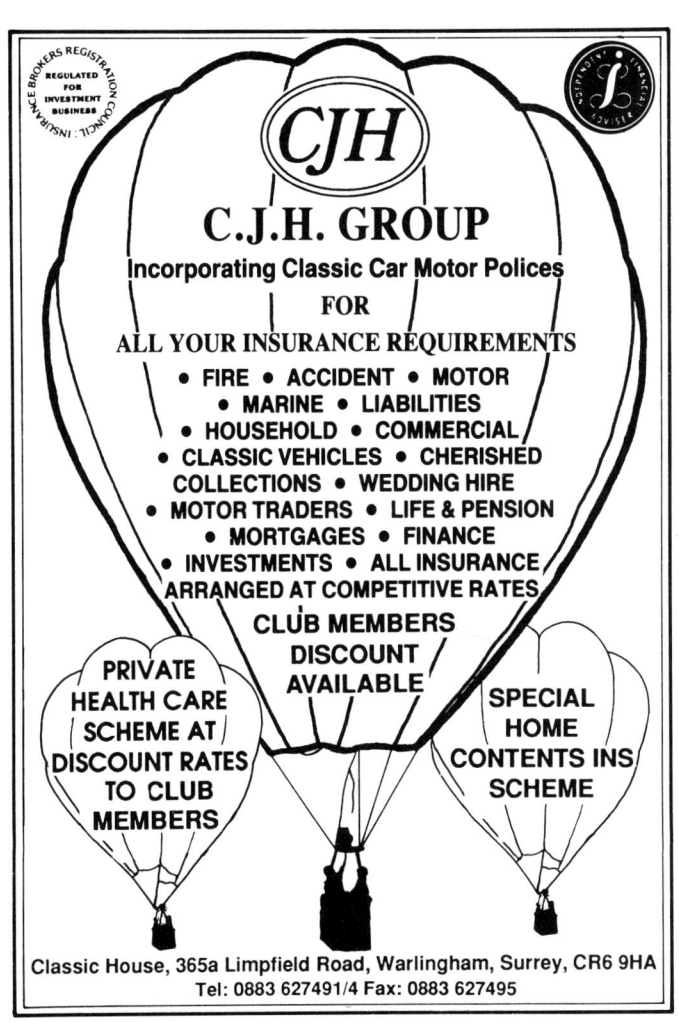

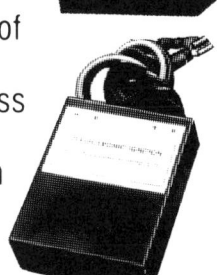

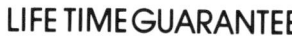

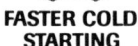

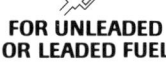

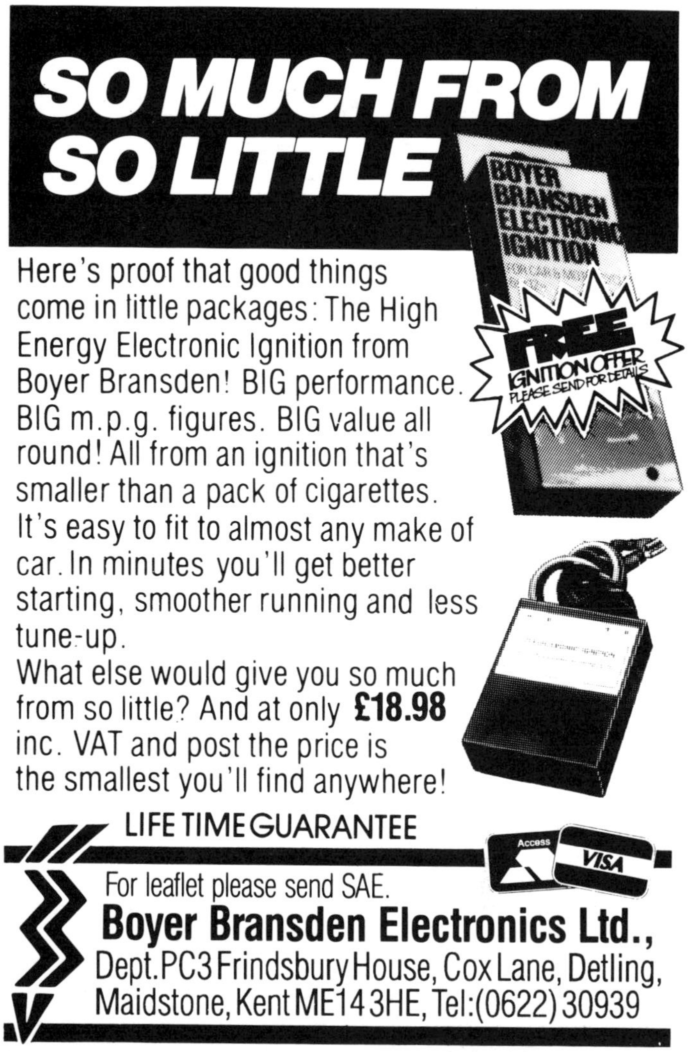

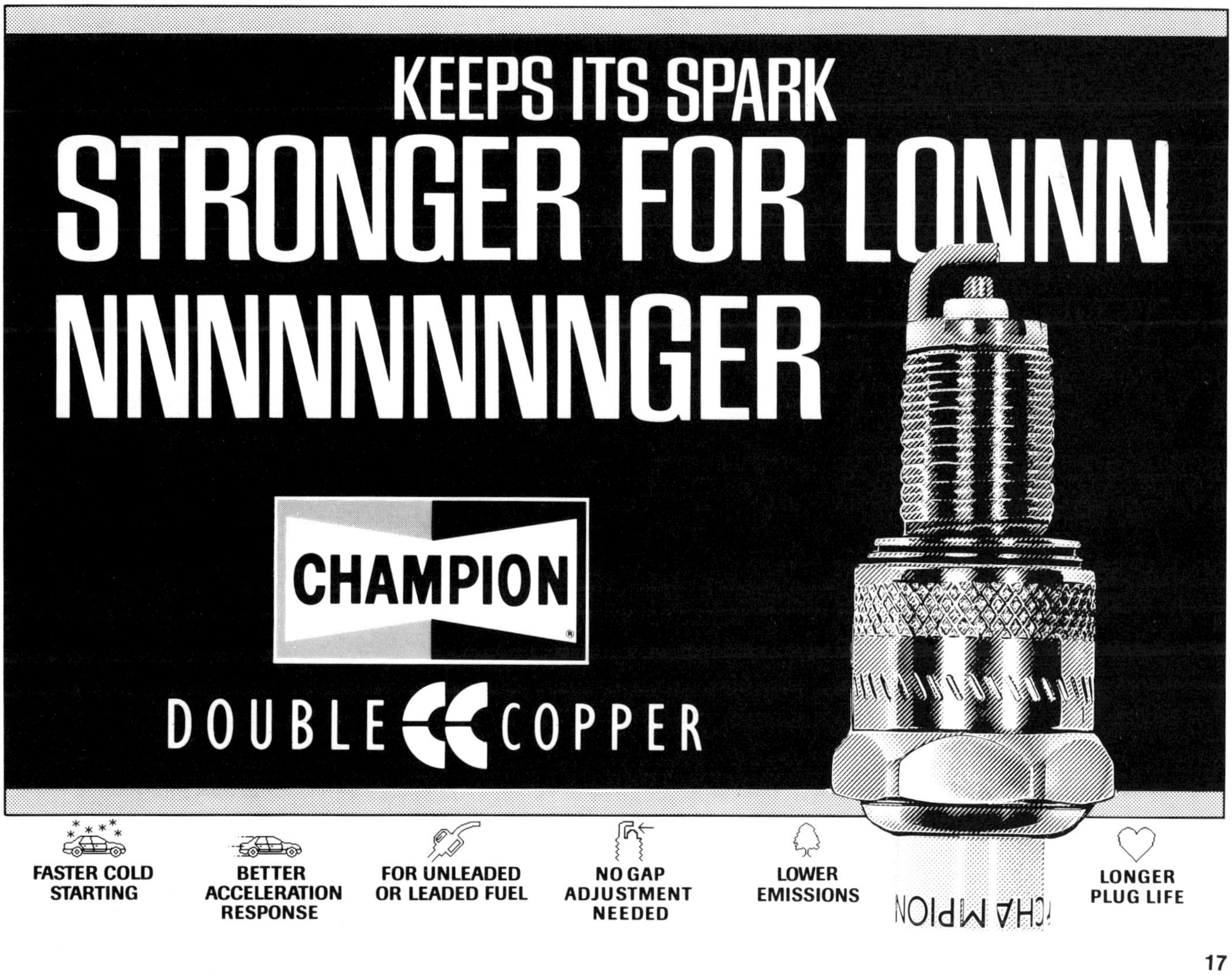

START RIGHT

Joss Joselyn explains the overhaul of a Lucas Starter Motor

LUCAS MODEL M35G & M35J

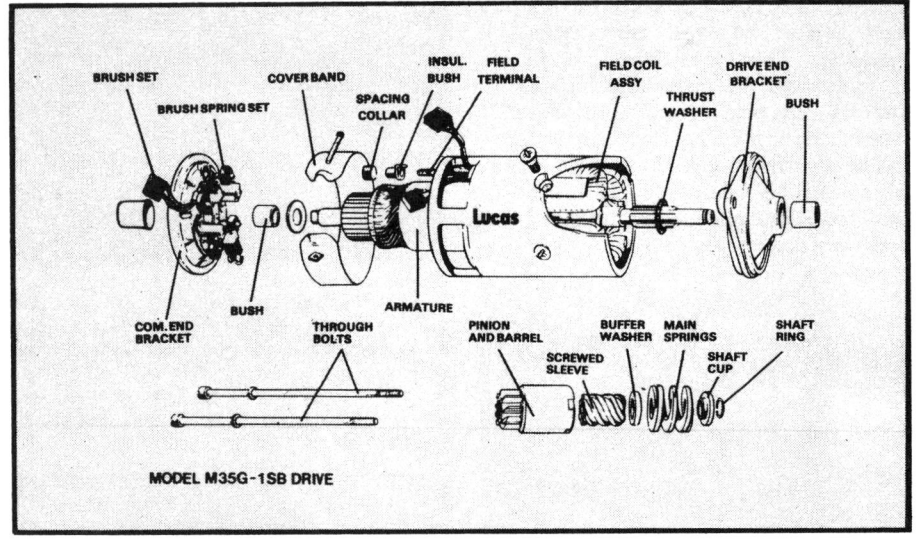

MODEL M35G-1SB DRIVE

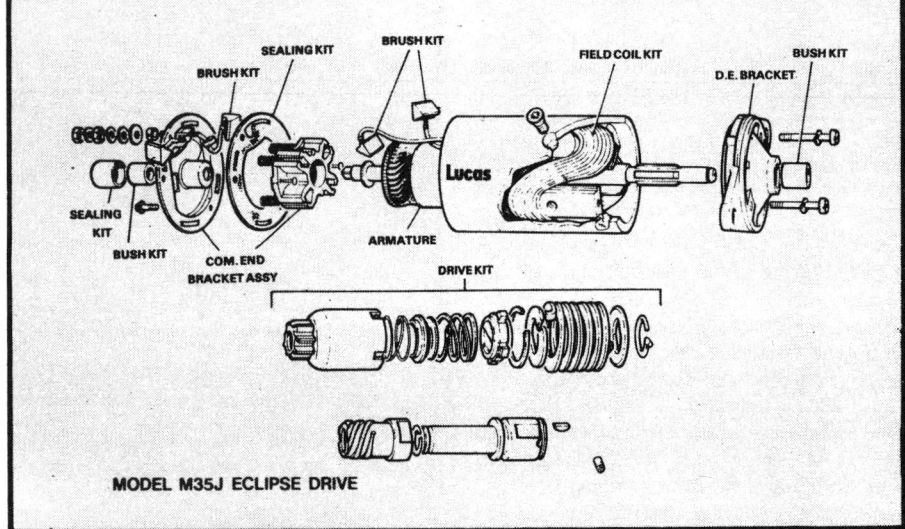

MODEL M35J ECLIPSE DRIVE

When you think about it, a car's starter motor leads a very strenuous life — perhaps your classic has a starting handle which will allow you to feel just what that starter motor is asked to do on a winter's morning. Another thing to consider is the lack of attention the starter gets, shoved into the depths of the engine bay and forgotten until something goes wrong.

You can be sure that the "something" will go wrong at the most inconvenient moment.

Once the possibility of faulty connections anywhere in the starter circuit or a defunct solenoid have been eliminated, anything wrong with the actual starter will have to be sorted out after the unit has been removed from the car.

The two starters being dealt with here were fitted to cars like the Triumph Herald, A.35 and Vauxhall Victor as well as a long list of others. Externally they appear similar except that the earlier M35G has a cover band over the brush windows while the later M35J does not. There are other internal differences, however, one of which is in the type of commutator fitted. The M35G has a cylindrical commutator while the later type

Tools You Will Need

Various spanners for removal. Screwdrivers. Soldering equipment. Small file. Narrow-nosed pliers. Vice. Bendix spring drive compressor. Hammer. Punch.

(Continued)

has a face or disc commutator and this means two entirely different brush arrangements, as will be seen from the photographs.

In the M35G, the field windings are insulated from the motor body (yoke) and the terminal post is incorporated with them. On the end plate the brushes are soldered directly to it.

In the M35J, the brushes are fully insulated from the end plate and joined directly to the terminal post. In this type the field windings are earthed to the yoke. These differences should become apparent when dismantling and details of this, together with overhaul and repair can be seen in the photographs.

Dismantling

The photographic sequence covers the M35G starter initially and the first basic steps in dismantling are simple and obvious. The armature is lifted out with the front (drive) end bracket still in position and this cannot be removed from the armature spindle without first dismantling the bendix drive. To free the rear (brush holder) plate completely from the yoke, the two field coil brushes will have to be lifted completely out of their holders.

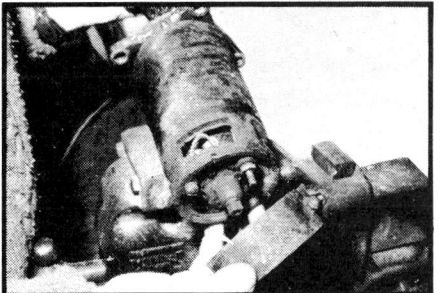

Dismantling is most easily tackled if the unit can be mounted in a bench vice and with the earlier M35G, the obvious first step is to remove the cover band to expose the brushes.

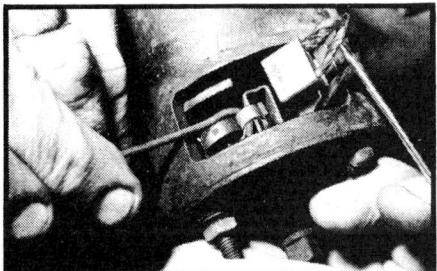

To make brush removal easier later on, hook the ends of the 'clock' springs off the back of the brushes and lift them out now. If the commutator is pulled out of the centre first and the brushes are allowed to move further inwards, they will be more difficult to remove. There are four of them incidentally.

Remove the nuts and washers from the terminal post and unscrew and pull out the two through bolts. Now the complete armature and front end plate can be pulled off in one direction and the end bracket carrying the brushes in the other.

Incidentally, if it is merely desired to inspect the brushes for wear and they do not, after all, require replacement, it will not be necessary to dismantle the unit — apart from releasing and sliding clear the cover band. The brushes can be hooked out with a piece of bent welding wire as shown in the photographs and inspected for length. If they are not badly worn, a piece of rag can be pushed in and the commutator rotated to clean it. Then the brushes can be checked for free movement in their guides and these latter cleaned if necessary. Then everything can be re-assembled and the cover band replaced. If, however, the brushes are worn and need renewal, the unit will have to be dismantled.

The brush length given by Lucas to indicate when renewal is necessary is 3/16 in. but most auto-electricians would fit new brushes if they were worn down beyond 3/8 in.

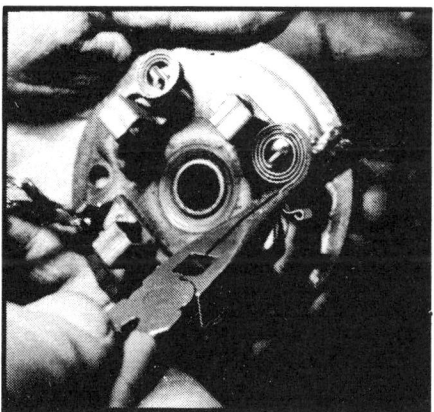

There is no real check on the flat wound 'clock' springs, except to note that when they are freed from the brush holder they should spring back between 90 and 180 deg. If they don't, fit new springs.

When fitting new brushes, both the earthed brushes on the end plate and the field coil brushes have to be soldered into position. The technique for this is described in the photographs and captions. Take care, when a new brush kit is purchased, not to mix them up. The end plate brushes have the shorter leads and the field coil brushes have the longer leads with the braided insulation covering.

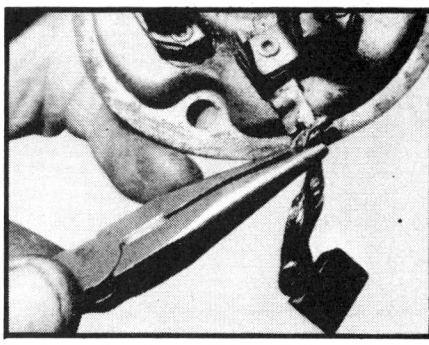

To change brushes on this model, use a pair of thin nosed pliers to unroll the soft metal terminal and then 'niggle' the braided lead to and fro until the solder connection is parted. If this proves difficult, use a soldering iron.

Before fitting the new brushes wash the brush holder in petrol or paraffin, and when fitting the new brushes into their guides, check that they are free to move. If they bind at all, relieve the sides very slightly with a fine file.

Tin the end of the pigtail of the new brush lead and use the long nosed pliers to roll it inside the copper tag. Solder into position finally using resin-cored solder and not Bakers fluid. A fairly hefty soldering iron will be best for this work. Take care to keep solder out of the lead and restrict it just to the end. Too much will spoil the flexibility of the lead.

The insulated brushes on the field coils are attached during manufacture by a special pressure welding technique. To change the brushes for new ones, the old leads must be cut about ¼in. from the joint. The pigtails that are left are then cleaned and tinned and the ends of the new brush leads are soldered to them.

Commutator

Wipe the surface clean with a petrol soaked rag and inspect for damage. Because the thickness of metal is not as much as with a dynamo, the amount of skimming that is possible is less. Usually, if the commutator won't clean up with emery cloth or glass paper, it is a case of fitting a new armature.

Because the segments on this commutator are not undercut, it is permissible to use emery paper to clean them up, provided the commutator is washed afterwards with petrol or cleaning fluid.

Clean up the commutator surface with a petrol-soaked rag and hopefully this should result in smooth shiny metal. If not use a strip of emery cloth or glass paper. If this fails, have it skimmed on a lathe by an auto-electrician.

Bendix Drive

If brush replacement and a general clean up is all that is necessary, there is no need to dismantle the bendix drive or to remove the front end plate. If the bendix is damaged, however, or the front bearing has to be renewed, dismantling cannot be avoided. A necessary aid for this is the little spring compressing tool shown. It is not expensive and it should be possible to purchase one locally. Alternatively, it can be purchased by post, along with any other starter parts required. Details are given separately on page 14.

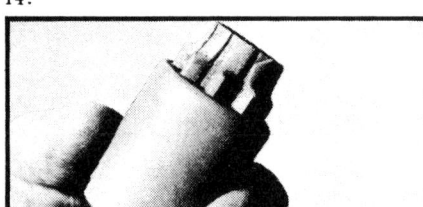

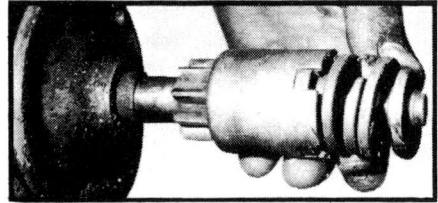

When it comes to the bendix assembly, here are two very obvious faults. One shows a badly chewed-up drive pinion and the other a broken main spring. Repairing either of these involves dismantling the assembly.

To dismantle a bendix drive a compressor for the spring must be used, as shown here.

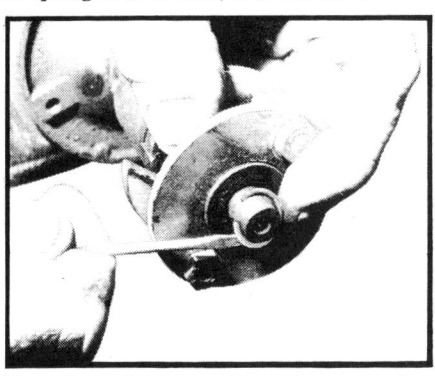

Once the spring is compressed far enough to clear the circlip, this can be fished out with a small screwdriver. It will jump, so take care not to lose it.

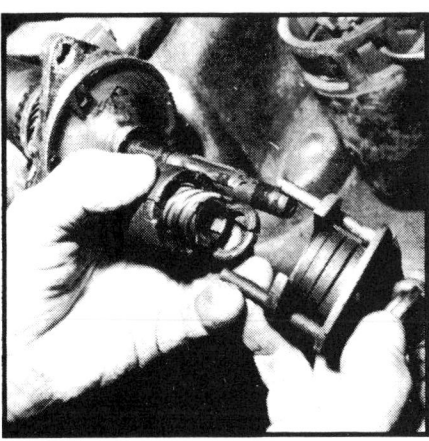

After this, the bendix assembly can be separated into its component parts.

Wash the various bits and pieces in petrol, cleaning fluid or paraffin. An old dish and paint brush are ideal for this job.

(Continued)

Bush Renewal

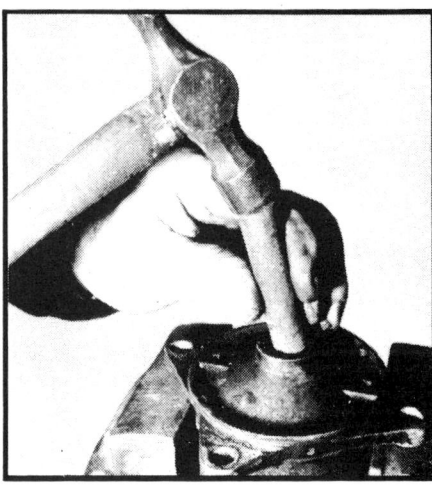

Bearings in this type of starter motor are all sintered bushes and if there is any discernible play, should be changed. Start by drifting out the old bush from the outside.

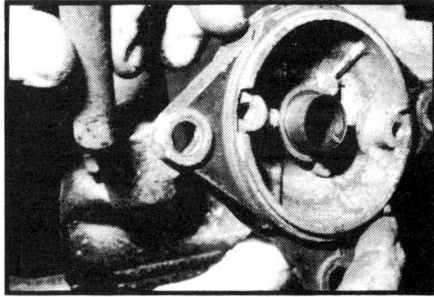

Here it is very nearly clear of its housing.

Soak the new bushes 24 hours in engine oil before fitting and then do the job using the vice as a press in conjunction with a suitable mandrel, as shown.

Later Model

Much the same points apply to the later M35J starter. The main point of difference is the commutator and brush arrangement. Even less cleaning up is possible with this disc type commutator and in most cases, significant

START RIGHT

(Continued)

damage means a new armature.

Because of the angular shape of the brushes and the type of guides they are housed in, cleaning up thoroughly is even more important if brush sticking is to be avoided. The method of fixing the field coil brush leads is the same as for the older model but there are two methods used for the end plate brushes. One is the same soldering technique used for the older model and the only point to watch with this is to note that one lead is longer than the other. Get them the right way round — it's annoying to have to solder them twice! The other brush fixing technique is very simple indeed. For this type the two brushes are supplied with their leads already attached to an exchange terminal post. The old brushes are removed together with the old post simply by taking off the fixing nuts and washer and are replaced the same way.

The only other significant difference is in the method of securing the rear (brush holder) end bearing. The later units have a small plate secured by two rivets. These are simply drilled out and new ones obtained at the same time as the replacement bearing. Otherwise the technique is the same as for the older model.

This is how the later M35J starter motor comes apart. Two screws are undone at the front (drive) end and undoing two smaller screws will release the back (brush carrier) end.

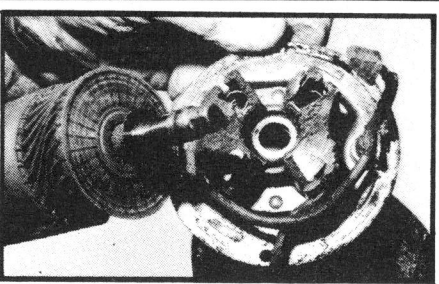

Here are the main components. Releasing the brush carrier from the yoke completely entails fishing the two angular-shaped field brushes from their boxes on the back plate.

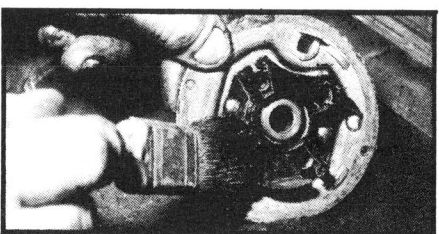

Cleaning methods are much the same and it is always worth doing this thoroughly to remove the usual accumulation of dust from the brushes.

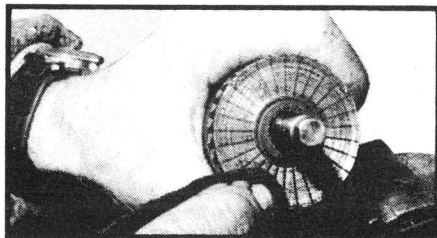

Here is the disc-type commutator found on this type of starter. If you are lucky, cleaning will only involve wiping with a petrol-soaked rag.

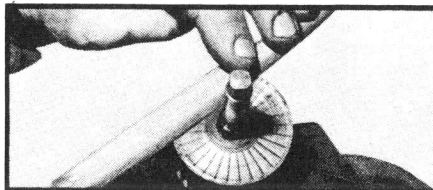

If it is more heavily damaged, provided the wear and scoring is not too deep, the face of the commutator can be cleaned up using abrasive paper wrapped around a flat block or a fine file.

Bearings on this type are the same as the earlier one, except at the brush carrier end, the bearing is held by a small plate secured with rivets. Renewal involves drilling these out and then refixing with new rivets.

Generally this is simply a reversal of dismantling but there are one or two points to note specially.

When reassembling the bendix, this is done dry. The spring compressor will have to be used again and care must be taken to ensure that the little circlip goes fully into its groove and that the end plate of the bendix assembly can pass properly over it. If the parts have been washed in paraffin, no lubrication at all is necessary as the washing will leave a very light oily film. If it has been washed in cleaning fluid, however, some auto-electricians consider it wise to use a very light smear of thin (3-in-1) oil on the screwed sleeve.

Note that if the pinion or the screwed sleeve are damaged both items will have to be renewed as a pair. The spring, however, can be changed separately.

The other point which is worth describing is the technique used to enable the commutator to be inserted between the brushes on the older M35G unit. All the brushes are assembled into their holders and when it is ensured they are free to move, they are temporarily fixed in the withdrawn position by locating the spring ends on the side of the brush instead of on the back.

When the unit is completely assembled, a piece of bent wire can be used through the access windows in the yoke, to re-position the end of each spring on the back of each brush, so pushing it into contact with the commutator.

Before re-positioning the cover band, ensure that there has been a time lapse, if petrol has been used to clean the inside. There must be no fumes trapped inside because running the motor will produce sparks from the brushes. It is not unknown for a newly reconditioned unit to explode the first time it is used!

One final point, if the unit was removed to change the bendix drive pinion, following its sticking in mesh, don't forget to inspect the teeth on the starter ring on the flywheel. If these are badly damaged too, more work will be required in taking out the flywheel and fitting a new toothed ring.   □

PARTS AND TYPICAL PRICES

Bendix spring compressor	£2.75
Pinion and sleeve (54252212)	£5.80
Pinion and sleeve for Ford (542420294)	£5.97
Pinion and sleeve for Triumph 1300 fwd (anticlockwise)	£11.20
Brushes for M35G	£1.22
Brushes for M35J	£1.95
Commutator end bush for M35G (250626)	20
Drive end bush for M35G (253316)	26
Bush kit (both ends) for M35J (54244773)	£1.09
Main spring	47

All prices plus VAT and postage. Surplus will be refunded.
(Available from: Shortlands Auto Spares, 87 Beckenham Lane, Bromley, Kent.)

Getting Wised up on Wipers

By Joss Joselyn

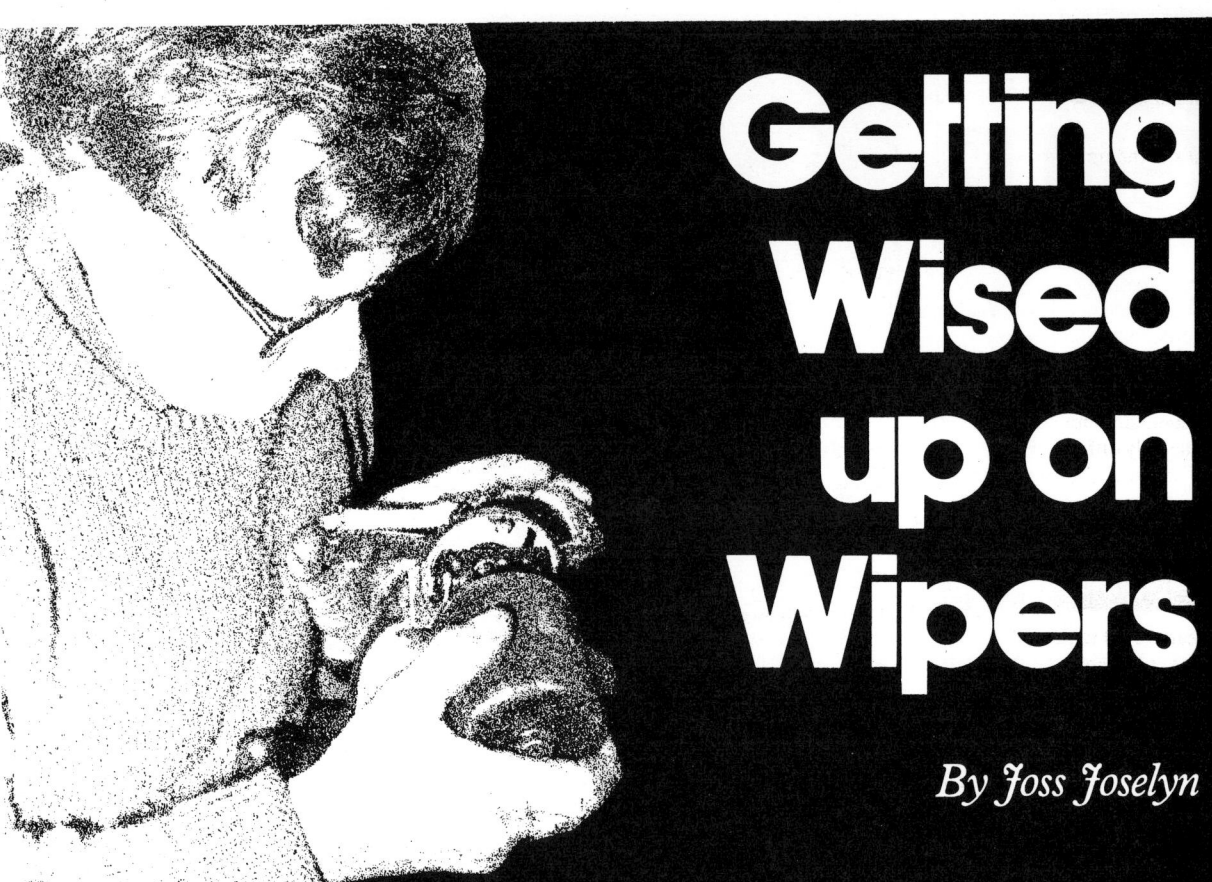

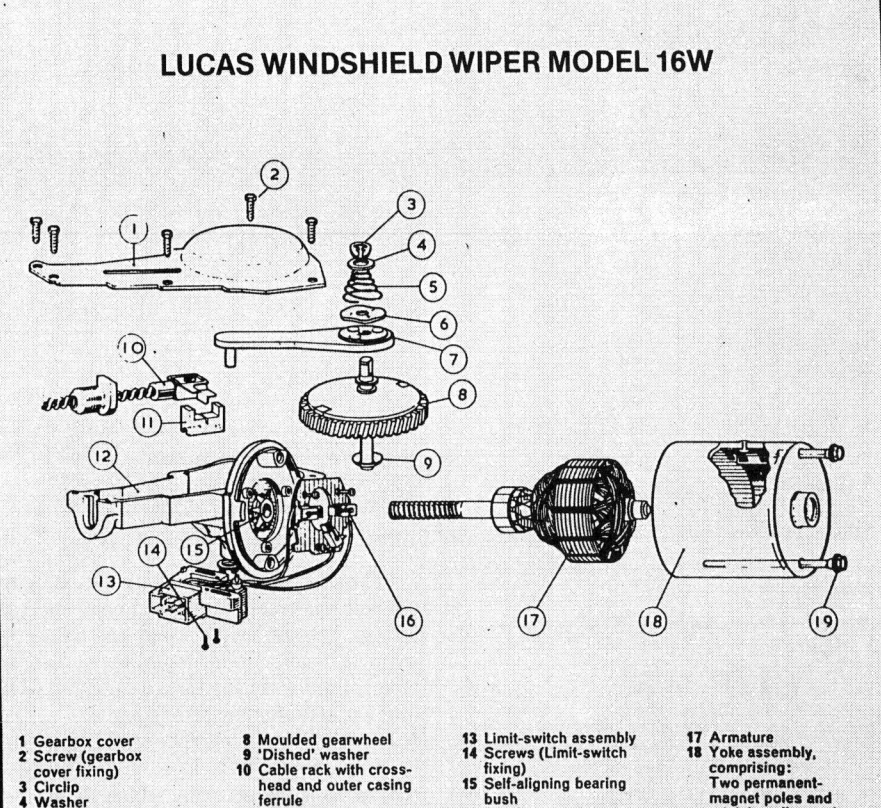

LUCAS WINDSHIELD WIPER MODEL 16W

1 Gearbox cover
2 Screw (gearbox cover fixing)
3 Circlip
4 Washer
5 Conical spring
6 Friction plate
7 Connecting rod assembly including eccentric cam

8 Moulded gearwheel
9 'Dished' washer
10 Cable rack with crosshead and outer casing ferrule
11 Moulded slider block (cable rack attachment) limit-switch operating
12 Gearbox

13 Limit-switch assembly
14 Screws (Limit-switch fixing)
15 Self-aligning bearing bush
16 Brushgear; comprising: insul plate and brush-boxes, brushes, springs, fixing screws

17 Armature
18 Yoke assembly, comprising: Two permanent-magnet poles and retaining clips and armature bearing bush
19 Bolts (yoke fixing)

Wipers are something most of us take for granted. We might change the blades occasionally, when they leave water behind as they sw eep across or start to make a funny noise, or something, but the hard working motor is usually left severely alone — until something goes wrong.

If you look back to "Staff Car Sagas" on page 68 of the July issue you can read about the run-in that Gordon Wright had with the wipers on his newly-acquired Triumph Herald. To be fair, he did sort his own bothers out before he took my name in vain in the last paragraph and said, in effect, "Joss will explain it all to you."

"No trouble," I said but I had reckoned without the machinations of Joe Lucas. Technically, there are no problems about the various motors and their gearing. Difficulty arises because Lucas have made various changes over the years and, as one motor grew out of another, there were overlaps. To make things as simple as possible, I'm only going to talk about the permanent magnet type units. I'm not going to confuse the issue by talking about part numbers because at one time these were all different,

Getting Wised up on Wipers

depending on what car the units were fitted to and then changed to a standard unit with different gears fitted, but even then the part numbering could only really be understood by a Lucas agent.

Positively my only reference to motor type numbers will be to say that we're starting with the standardised 14W or 16W, which is the type shown in most of the accompanying photographs. There are single and two-speed versions of this, incidentally, but they both use three brushes and the only difference is that the single-speed does not use the third brush, (yellow wire on two speed.)

REMOVING THE MOTOR

If the motor is connected to the wipers via a rack and wheelbox mechanism, the wiper motor/gearing unit can be taken out of the car in one of two ways. Possibly the simplest is to unbolt the gearbox cover and disconnect the cable rack by taking out the connecting rod. Before the rack can be cleared from the gearbox, however, the wipers themselves will have to be removed so the splined drives can rotate.

A second method is to take off the motor and gearbox unit together with the rack complete. Again the wiper blades should be taken off first and then the big nut on the end of the wiper gearbox, which holds the rigid rack tubing in position, should be undone. The wiring connections are pulled off next at their patent Lucar connectors and then the mounting nuts can be undone and the unit removed complete.

Units with a linkage instead of a rack and wheelbox drive, often have to be removed complete with linkage — an operation which might involve dismantling half the dash panel and facia.

In the case of a rack drive, it is better to take this off with the motor and gearbox so that it too can be inspected for wear. At the same time, the wheelbox spindles should be checked to see if they are free.

If there is evidence of wear and this is often shown by wiper arms which travel too far and thump into the edge of the screen, a cheaper expedient to renewing the wheel boxes is to turn the wheels through 180 deg. so that an unworn section engages with the rack.

DISMANTLING

Most of the major steps can be seen in the photographs. Four screws hold the cover and once this has been removed, the connecting rod can be lifted off and the rack disconnected. The connecting rod is held by a circlip, which must be carefully pushed off. Check the bearings in the connecting rod to ensure they are not binding; they should not be excessively worn either.

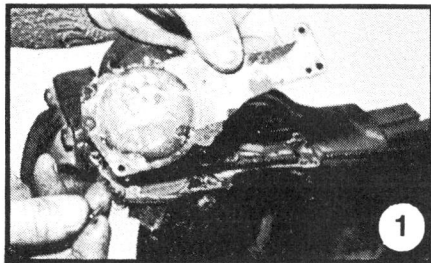
Dismantling begins by removing four screws and lifting off the cover.

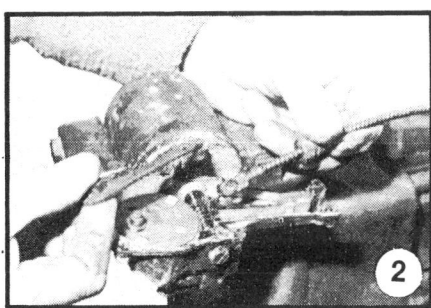
The connecting rod is lifted out, disconnecting the rack gearing.

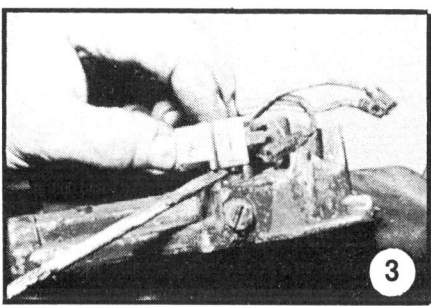
Removing the switch.

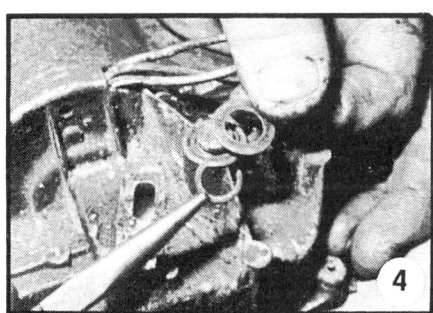
Gear spindle circlip and washer being removed.

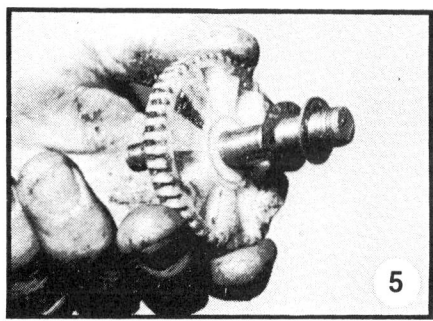
Note the two washers on the spindle. Shaped one goes concave side towards the gear.

Note the position of the leads on the plastic switch — a small sketch is the best bet — and then pull them off their terminals. Lever up the outer end of the switch with a screwdriver to disengage the locating peg and slide it out.

To take the gear out, a circlip and washer have to be removed on the outside of the box first. If the gear seems reluctant to move, it could be due to a slight burr on the spindle. If this is removed with a fine file, it should then lift out. Be careful not to lose the dished and flat washers on the spindle; they could be hidden in the grease, so don't miss them. The curved washer goes next to the gear convex side outwards. Its job is to take up the end float, acting as a sort of spring.

Before taking the body off, note the matching marks — an arrow head on the gearbox housing and a line on the motor casing. It is important because, if you get them fitted together a half turn out, the motor will run backwards.

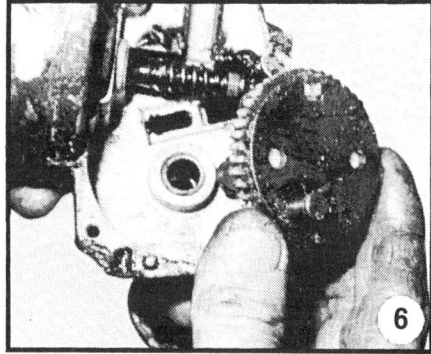
Lifting out the gear.

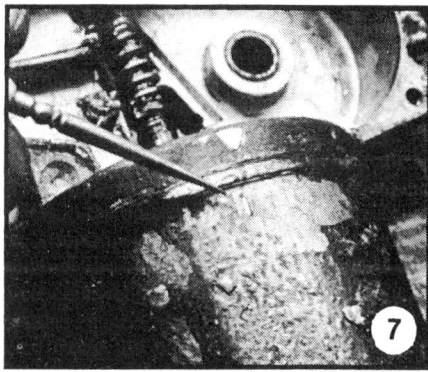
Alignment marks on the motor casing.

The armature comes off with the casing and if possible is best left there and not dismantled further, because taking it out can reduce the magnetism and cut down on the power of the motor. Problems can also arise because of the little ball at the bottom of the spindle.

In the motor photographed, it is captive but in some motors it is loose and difficult to replace, as is the little disc which sits inside a recess in the bottom of the body. The reason is the strong magnetic pull to either side. Once down at the bottom, it has to be pushed into position using a non-magnetic tool — plastic or wood.

There is not normally a lot of wear on the commutator and it can usually be rubbed clean with fine glasspaper, without removing it from the cover. If there is more wear than this, it may be possible to remove it by skimming on a lathe; but there is not a great thickness of metal and an auto-electrician is best consulted on this point. It does, of course, mean removing the armature.

The brush gear on the later unit shown in the photographs can be changed as an assembly — plate and three brushes — merely by undoing three screws. At one time the job involved soldering but even on older motors this brush assembly is now substituted and soldering is eliminated.

Pulling off the motor end plate to reveal the brush assembly.

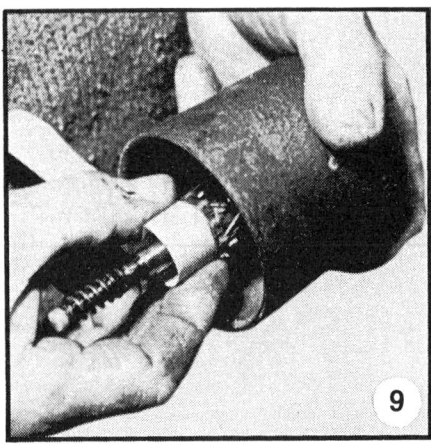

Use fine sandpaper to clean up the commutator.

Ball at the bottom of the spindle (captive on this motor).

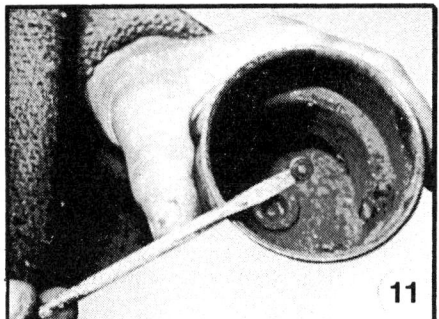

The little packing washer at the bottom of the motor cover can be difficult to refit because of strong magnetic attraction to the sides.

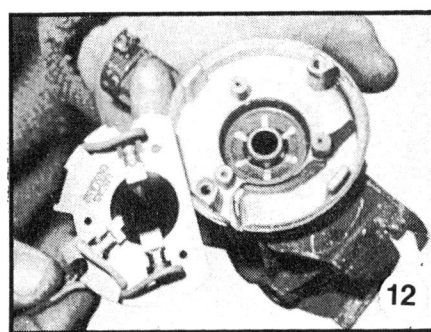

Removing three screws releases the brush assembly.

CLEANING, INSPECTION AND ASSEMBLY

All the parts can be cleaned up using cleaning fluid and an old paint brush. Make sure the main bearing which is set in a spring housing (Fig. 12) is a tight fit in the motor end cover and not flapping around loose. It must be firm in its spring collar and not rusty. Lubricate the bearings with light oil.

Check the teeth of the gear wheel for wear. If there are any signs of damage, get a new one. If the right gear is not obtainable, buy any one the correct size and change the metal plate and spindle for the old one. It simply taps through. The thing to notice is the relationship between the driving pin and the switch cam on the other side, although the cam, incidentally, can also be changed over; it's a simple push fit in holes in the plate. If you get this wrong, parking will be 180deg. out.

When installing, grease the gear wheel assembly generously — HMP grease should be all right for this.

Reassembly is a simple reversal of the dismantling procedure, except perhaps for getting the brushes to clear the commutator. This is a bit tricky and the way to do it is to use an old electrical screwdriver inserted between the rim of the body and the end plate and fiddle them up on to the commutator one at a time. It gets a bit frustrating when you get two of them in plate and the third one slips, but it can be done.

PARKING MECHANISMS

With the later type of motor, the parking of the wipers is automatically sorted out by the

Getting Wised up on Wipers

cam or ramp on the back of the gear wheel bearing on the little plunger in the switch (see Fig. 16). The way this works is not too difficult to understand. The first part of the plunger's movement breaks two contacts in the main electrical feed to switch off the motor. The second half of the movement closes two contacts which effectively short circuits the armature, which then gives regenerative braking. That was the sort of electrical braking system they used to use to stop trolley buses!

A lot of trouble can be caused with this switch if one of the internal switch blades breaks. This might stop the motor working or it might blow the fuse but the answer is always to fit a new switch to the motor. With the later motor, this can be fitted without any dismantling but on earlier 14W's, things had to come apart because the fixing screws for the switch were under the main gear and to get at them, it had to come out.

The older type motor has a different type of parking switch. This is a sort of hybrid in which, although it has the later permanent magnet motor, it retains a previous parking switch. This switch was the one fitted to Gordon Wright's Triumph Herald and can be seen in Fig. 3 and 14. It consists of a spring-loaded finger moving around a contact disc with an insulated section. When the wipers are switched off at the panel switch, an electrical feed is continued through the parking switch, until the finger reaches the insulated section, when the power supply is cut and the motor stops.

The way to adjust this, so the blades park in the correct spot, is first to slacken the screw holding the gearbox cover. Then, with the ignition switched on and the wipers switched off, rotate the switch cover. As you do this watch the wiper blades — they will move across the screen towards the parking position. Stop when the right spot is reached and lock up the cover.

All this, however, should not be necessary if, before dismantling this type of motor, you take one simple precaution. Note the position of the pip on the switch cover and how it lines up with

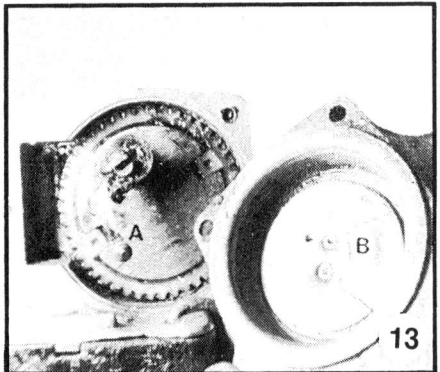

The older type parking arrangement. A is the spring-loaded sweep contact and B is the insulated segment on the disc.

Getting Wised up on Wipers

one of the marks on the body. If there are no marks, scratch matching lines across cover and body before dismantling. If they are matched up properly, the parking will be correct.

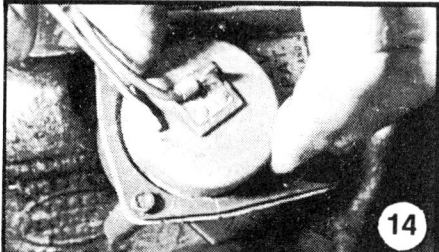

Positioning marks on the parking switch cover.

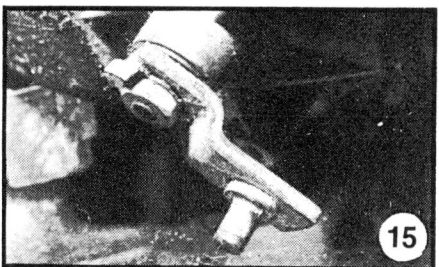

This is the cranked drive end on a unit of the type with no connecting rod.

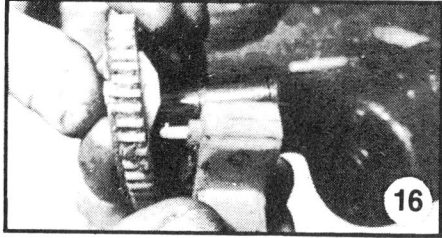

How the cam or ramp on the back of the gear operates the plunger in the switch to switch off the motor and park the blades.

BUYING BITS

From Gordon Wright's tale of kerbside enterprise in the July issue you will have gathered that an exchange motor will cost in the region of £27, compared with £31 for a new one. Don't forget also the possibility of finding what you need at a breakers yard (sorry, vehicle dismantlers). The best bet here is to take your old one with you so you can get an exact comparison between old and new. Check particularly the shape of the motor housing to make sure it is either a permanent-magnet type or a wire-wound-field type — whichever your old one was. Check the method of switching — separate switch or gearbox-end-cover switch and the rest of the motor for identical fitting, type of drive, etc. From the two, you should be able to build one good motor without too much trouble and don't forget it is still possible to buy new bits,

like the following:-

Armature	54709161	£10.70
Brushgear	54704696	£3.07
Gear (Av.size)	57702587	£5.69

(may vary with motor).

Switch	54705657	£3.65

All these prices are without VAT and were supplied to us by Shortlands Auto Spares, 87 Beckenham Lane, Shortlands, Bromley, Kent. If readers cannot obtain these from their local Lucas agent, they will be pleased to supply but prices will, of course, be subject also to postage and packing.

There is one other variation on the Lucas windscreen wipers of this era which has not been mentioned so far. This is the type designed to be used with a linkage instead of a rack. In this unit the motor is the same as that already described and so is the switching — the sweep contact on a disc. What is different is the gearing. Instead of a single gear, there are two, designed to reduce the speed of rotation. The motion at the end of the gearbox is still a rotary one, instead of a reciprocating push-pull movement when there is a rack. The addition of a rotary crankpin then supplies the necessary movement to the linkage to the wiper spindles.

The only difference with this unit is the second gear wheel and the fact there is no connecting rod. The overhaul procedure is exactly the same in principle.

CONTINUED FROM PAGE 16

The Generation Game

Changing them is simply a matter of a single screw through the tag on the end of the lead. While the brushes are removed, clean out the inside of the holders or guides and then when the new ones are installed try them for free movement. If they tend to bind a little, use a fine file just ro relieve the sides until they are completely free.

REASSEMBLY

This should cause no problems, provided the order of washers, spacers and other small parts was noted when dismantling. Reassemble the pulley and front end bracket to the armature spindle first. Carefully drive the end bracket on using a hammer and a piece of tubing as a mandrel. Insert the Woodruff key in the shaft after sliding on the spacer. The pulley goes on next, followed by the spring washer and nut.

Slide the armature and front end bracket assembly into the body, rotating it until the little moulded pip on the end plate locks into the recess in the rim of the body.

Hook the tail of the brush spring off the back of the brush and let it bear on the side,

with the brush drawn back up into its guide. The side pressure of the spring will hold the brush firmly in that position so it will not foul the commutator when assembling. When both are withdrawn in this fashion, fit the rear end bracket and reinsert the two through bolts.

Now, the final job is to fish through the windows on the rear end bracket with a screwdriver and hook the spring up from the side of the brush onto the back so that it pushes it down onto the commutator. Repeat for the second brush.

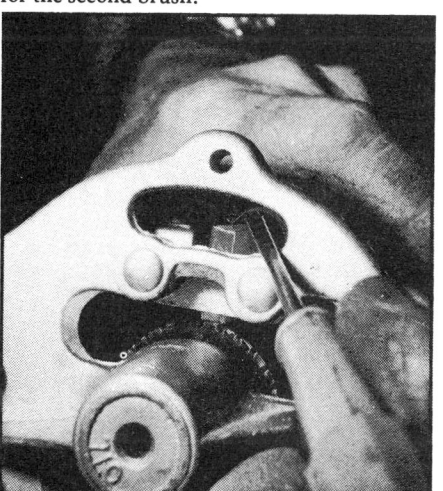

After reassembly this is how the tip of the spring is re-located on the back of the brush to bring it into contact with the commutator.

Before refitting into the car, check the condition of the fanbelt and get a new one if necessary. Ensure that the tension is correct before finally tightening the mounting bolts. An over-tensioned belt will cause rapid wear of the dynamo bearings and in most cases the water pump bearings as well. In use the rear brush should be lubricated with light oil from time to time.

TESTING

Test that the dynamo is charging by using a short piece of wire to connect the F terminal and the D terminal together. To the centre of this connect the negative lead from a voltmeter and clip the other lead to earth on the dynamo body. Run the engine at tickover speed and there should be an instrument reading of around 15V.

WIPER MOTOR STRIP

In our September issue 1980 issue we looked at the Lucas Model 16W; here Joss Joselyn examines some of the earlier types.

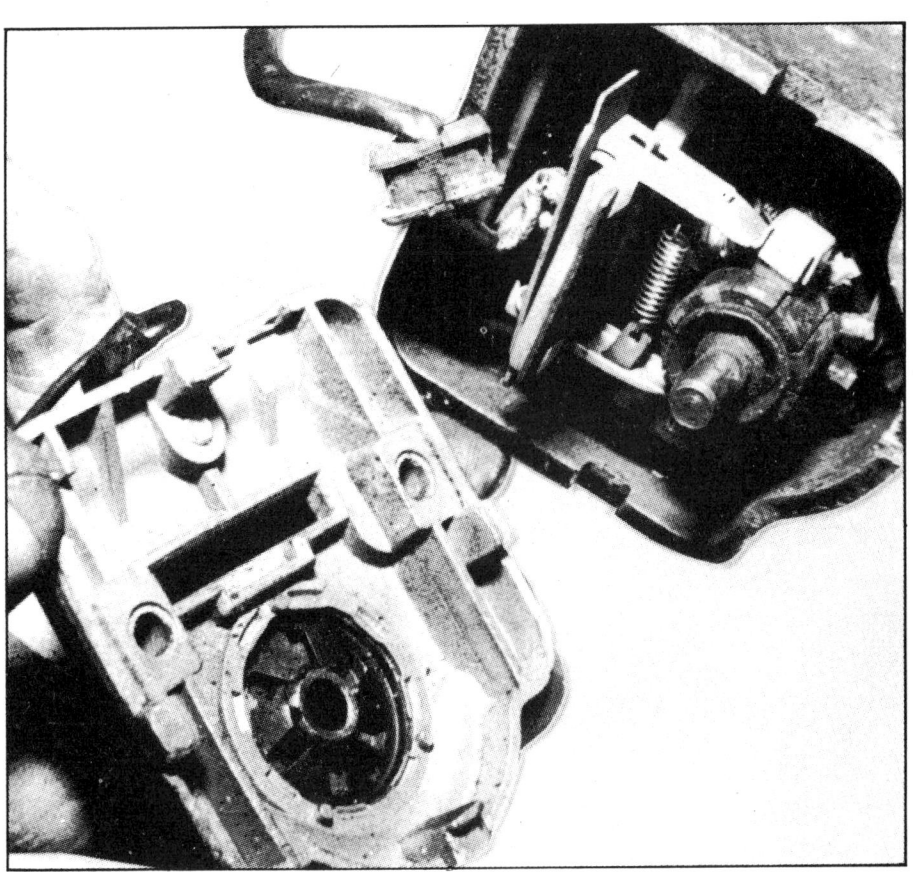

People who run new cars can perhaps be forgiven for expecting that their windscreen wipers will work every time they prod the button. Those of us with older models, however, cannot really expect the wipers to go on for ever; eventually something will go wrong. Rather than have this happen at the most awkward moment (night time when it's hissing down), it's not a bad idea to have the motor and its linkage out from where it lurks and check it over.

Tools you will need

BA Spanners 2 x 4 • Small pointed pliers • Screwdriver • Small hammer

The motor we are talking about here is the older Lucas type — the one with the square case. There were a few modifications to this over the years before it changed completely to a permanent magnet motor in a round case, but the differences are only minor. The one shown in our photographs actually came out of a Mk. I Cortina but it's basically the same as lots of others of the same period.

ELECTRICAL CHECK

You'll need to know first that the problem is actually in the motor before you start taking the trouble to remove it. Before you do anything else, visually inspect the wiring and connections that you can see, looking for

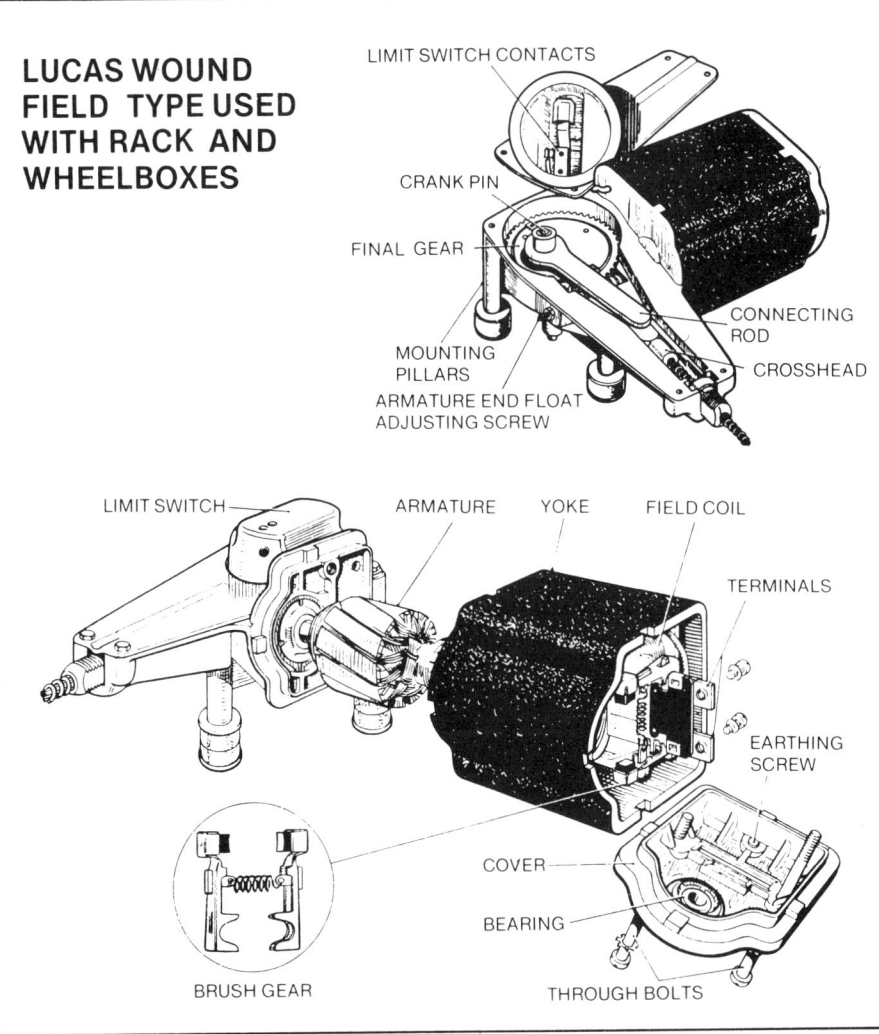

LUCAS WOUND FIELD TYPE USED WITH RACK AND WHEELBOXES

LIMIT SWITCH CONTACTS
CRANK PIN
FINAL GEAR
CONNECTING ROD
CROSSHEAD
MOUNTING PILLARS
ARMATURE END FLOAT ADJUSTING SCREW

LIMIT SWITCH
ARMATURE
YOKE
FIELD COIL
TERMINALS
EARTHING SCREW
COVER
BEARING
THROUGH BOLTS
BRUSH GEAR

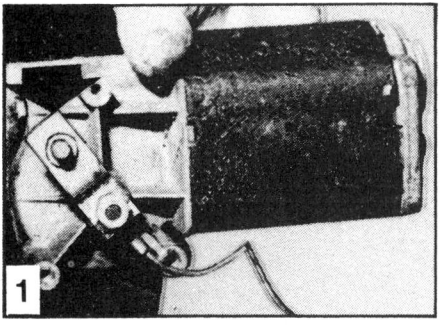

This is the motor shown taken to pieces in the following photo-sequence. It's of Lucas origin and was fitted in the Cortina Mk I. The actual motor part is identical to others fitted in other makes of car at the same time but there is a difference in the gearbox. Note that the reciprocating action here is achieved by means of a rotating arm on the outside of the gearbox (arrowed).

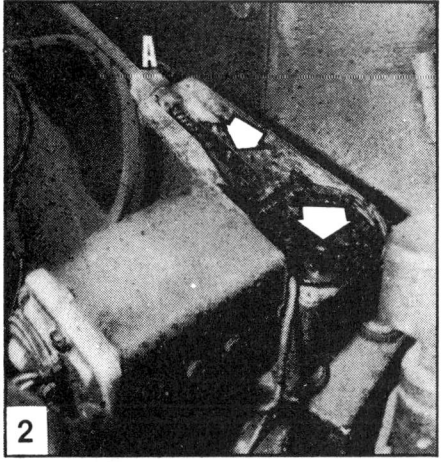

The other type of gearbox (fitted to BMC cars for instance) is shown here. The cover has been taken off and the crank removed to show the reciprocating mechanism. The crank fits between the points arrowed. To remove this type of Lucas motor and gearbox, the crank and cover have to be removed as already shown. If then, in addition, the union marked 'A' is undone and the mounting bolts released, the whole assembly can be lifted out of the car.

Locating pegs in the rim of the commutator end plate fit into cutouts in the edge of the body. It is released by taking out the two long through bolts. This will reveal the simple 'crocodile arm' brush assembly as shown.

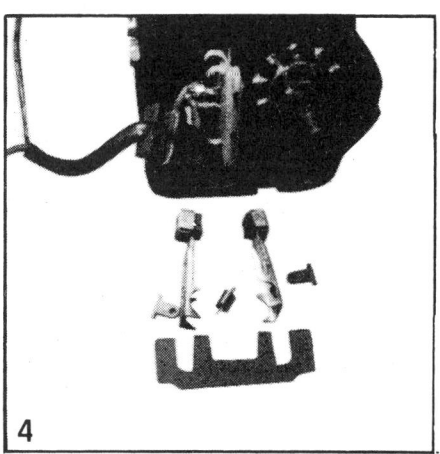

Here are the components involved. The brush arms are a location fit, held in contact with the commutator by spring tension and keyed by means of the small fibre locking plate.

Remove the body and pull out the armature so that the condition of the commutator can be inspected.

corrosion, broken wires or loose terminals.

If all appears well, further checks can be made using a simple test lamp. Find out first if current is actually reaching the motor. Do this by connecting one side of the test lamp to the feed terminal on the motor and the other to earth. Switch on the ignition and the motor and the bulb should light.

If there is no light, check back to see if there is current at the switch. If there is, there must be a break between switch and motor. If the bulb still doesn't light, check back further through the circuit to ignition switch, fuse box, etc.

If the first check at the motor lights the bulb but the motor does not operate, it must be the motor itself or the earth connection that is faulty. Sometimes if the motor is in a convenient position and access is easy some work can be done on it in situ, generally, however, the best plan is to remove it from the car.

What this involves will vary from model to model. With the rack and wheelbox type, used by BMC among others, the top should be removed from the gearbox. Then the connecting rod is released by taking off the small circlip at the gearwheel end and lifting it clear. The gearbox can be separated from the rack by undoing the union nut on the rack tube at the end of the gearbox and lifting out the rack and its crosshead. The motor itself can then be removed by pulling off the electrical connections and releasing the motor mounting bolts.

With the type of motor that employs a linkage drive, motor and drive usually have to be removed together and then separated on the bench by disconnecting at the drive arm on the motor spindle. With this type, it is a matter of removing the wiper blades first, followed by the securing nuts, washers etc. Then the motor itself is disconnected electrically and disengaged from its mounting and the whole assembly lifted out of the car.

Where the wipers are mounted in the engine compartment on the bulkhead, it is reasonably easy. When they are situated under the dash panel, it may mean dismantling half the fascia.

Once the linkage type is on the bench, mark the position of the linkage in relation to the motor drive before dismantling. If the unit has been removed starting with the blades in the park position, the self-park arrangement will not have been upset.

Similarly with the rack and wheelbox type, mark the relationship of the switch cover with the gearbox before removing it. There is actually a pip marking in the cover which can be noted but it does no harm to put on extra scribe marks as well.

BRUSH RENEWAL

This is the most likely job that will need doing and with this motor it is not difficult to obtain a new set of brushes from an auto-electrician or from a Lucas agent. Start by mounting the motor in a vice and taking out the two through bolts from the back. The end plate can then be tapped free and lifted off.

The brush arrangement is now easily visible
(Continued)

WIPER MOTOR STRIP

(Continued)

and it is a simple matter to check if the brushes are worn or not. If they have worn down to a length of 3/16in. or less, fit new ones. In really severe cases of wear I have seen the brushes worn away completely and half the metal holders as well; the motor had only stopped working when the brush holder had parted company with its arm.

Another possible fault here is that the link spring breaks and the brushes lose effective contact with the commutator. Take out the brushes by first removing the fibre locking plate and then just lifting them free.

Separate the armature from the motor body and the body from the gearbox and clean all the parts in a bath of cleaning fluid or petrol. Examine all the internal wiring for signs of damage or burning. Simple breaks or insulation damage can be repaired but serious burning or damage will mean a replacement motor.

Clean the commutator with a petrol soaked rag and if this does not bring it back to bright metal, use fine glasspaper. If this does not eliminate damage, i.e. there is pitting or severe wear, a new motor will be needed.

It is possible to renew the armature bearings but whether it is worth while is something you'll have to make up your own mind about. They have to be chipped and cut out initially using a small sharp chisel and taking great care to avoid damaging the end plate. New bearings need soaking for 24 hours in 20 SAE oil before fitting. Bear in mind that a worn bearing bush probably means that the armature spindle is also worn and it is possible that the most economical way out of the trouble is to fit an exchange motor.

Motor reassembly is simply the dismantling

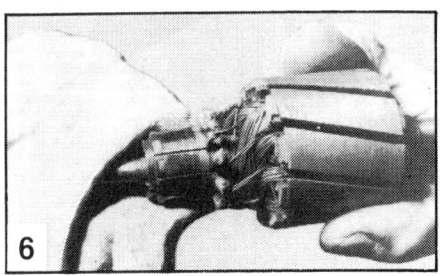

6

If it is in good condition, it can be restored to a clean, bright, smooth surface simply by cleaning with a petrol-soaked rag or at most with a piece of fine glasspaper. The connections from the windings should also be inspected. If there are signs of melted solder, a replacement motor is the best bet.

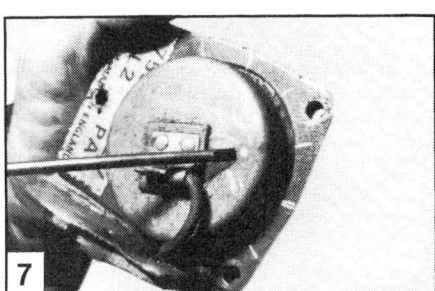

7

Three screws hold the switch cover (four on the rack and wheelbox type). Before removing these, note the position of the little 'pip' on the cover and ensure correct replacement by making marks across on to the gearbox. Correct blade parking depends on its position.

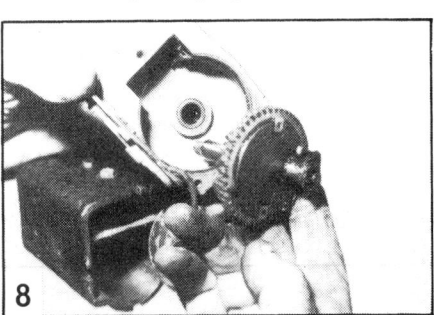

8

The Nylon-toothed gearwheel can be removed after taking off the circlip from the end of its spindle on the Lucas rack and wheelbox type. Here it lifts out together with its self-park contact.

procedure reversed. Make sure however, that the armature bearings are oiled and the self-aligning bearing at the gearbox end should be immersed if possible in oil for several hours prior to re-building.

Check the movement of the new brushes in their holders and if they show any tendency to stick, gently relieve the faces using a small fine file.

THE GEARBOX

With both types of gearbox it is possible to change the gearwheel. The type which operates the rack and wheelbox drive has a cupped washer underneath — take care not to lose it and before pulling the shaft out, make sure there are no burrs on it which could damage the bearing. This item is held in place by means of a circlip.

All you can do with it is inspect the teeth for wear. If you find any, fit a new gear, making sure when you buy it that you get the right one. It can make a difference to the parking position of the wipers, and to the angle of wipe.

Pack the whole of the gearbox with clean grease when reassembling. Ensure that the marks you made earlier lining up the domed switch cover are correct, and the parking position should be right.

As usual, everything depends on what you find when you look inside the motor. If it's just a case of new brushes and a commutator clean up, combine this with a complete clean up of the inside and regrease the gearbox. This can save the cost of a new motor.

If you find, however, that the brushes are worn, the commutator is worn or damaged, there's a lot of shake in the armature spindle, the gearwheel has worn teeth and the whole inside is filthy with carbon dust, it will be quicker, simpler and cheaper in the long run to fit a new motor.

NEXT MONTH
Rover 6-cylinder engine strip down.

Check-out for an ACR

Now that cars with alternators are becoming classics, Peter Wallage takes the lid off the Lucas ACR range

Quite a few cars with alternators are now moving into the classic range and owners are finding the charging system very different from the old dynamo and regulator box they were used to. The handbooks and manuals are often not a lot of good. Some say that alternators don't need servicing and leave it at that. Others tend to skirt round the subject and tell you to take the car to your local dealer for a check without telling you that he'll be pleased to relieve you of anything from £40 to £100 for an exchange unit when the chances are that yours can be put right for anything between a fiver and twenty quid.

The most popular range of alternators is the Lucas ACR range, still current, which doesn't need an external regulator. You might find cars from the early days of alternators fitted with the Lucas AC range, usually a 10AC or 11AC, which did use an external regulator, but these are more difficult to service and not so efficient as the ACRs; parts are becoming more difficult to find so, unless you're a stickler for originality, you'd be better off swapping to an ACR which you can get from any breaker's quite easily. Make sure you get the later 'machine sensing' model because the early ACR battery-sensing models weren't so reliable. You can spot the difference straight away because the later machine-sensing models have the so-called European termination, using only one connecting block with three push-on Lucar connecters in it. The wiring to the battery is quite simple.

Some handbooks say you can change the brushes on an alternator without taking it off the car but, if it's awkward to get off, it's going to be even more difficult to get at the tiny screws holding the brushes, so unship it and take it to the bench where you can see what you're doing and also clean the slip ring and the connecters.

Unless the front bearing's worn, usually caused by mounting the alternator out of line with the other pulleys, or straining the fan belt far too tight, you probably won't have to take the pulley off or take out the long through-bolts that hold the main body parts together.

Start by taking off the black plastic cover which is held by two ¼ AF bolts. They're shrouded quite deep in the cover, so you'll

You can take the brushes out to inspect them without any further dismantling.

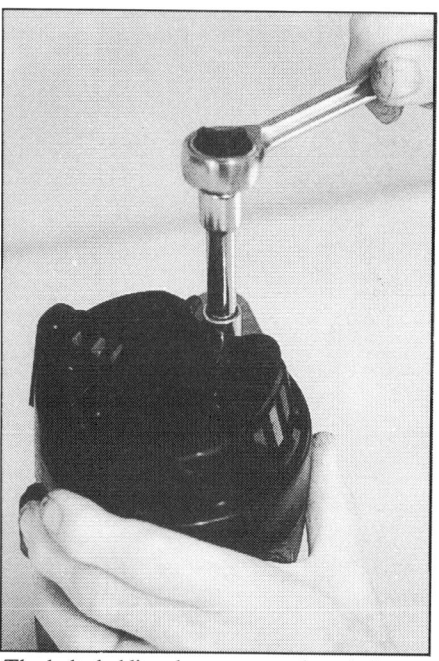

The bolts holding the cover are shrouded so you'll need a small socket or box spanner.

This is the regulator box. The later type of 14TR regulator doesn't need a surge protection diode.

need a small socket or box spanner. Under this you'll find the diode rectifier pack on to which the connecter plugs, a fairly large white nylon brush housing and, attached to this, a square aluminium box which is the regulator.

On ACR alternators built before 1980 or thereabouts you might also find a separate diode, looking a bit like a small capacitor out of a distributor. It sits alongside the rectifier unit, on its right when you're holding the alternator with the connecters at the top. This is the surge protection diode, put there to protect the regulator should someone remove a battery terminal with the engine running. A short-circuited surge protection diode is quite likely the culprit if the ignition warning light stays on when you rev up the engine and you've checked the obvious things like dirty connections and fan belt tension, though a faulty regulator can also cause the ignition warning light to stay on and, in this case, the warning light sometimes gets brighter as you rev up. Later ACR alternators don't have a surge protection diode as the regulator was improved and didn't need it. If you fit one of the later regulators you can take the surge protection diode off, though if it's in good order you can leave it on as extra protection.

All ACR alternators built for the past nine or ten years are fitted with the improved 14TR type regulator and, when Lucas first introduced it, they either painted it gold or put a gold band on it to indicate that it didn't need a surge protection diode. However, as the older models got fewer and fewer in Lucas service centres, they dropped

the gold paint and went back to a plain aluminium box. The only way to tell if a plain 14TR is an improved model or not is by the part number, but your local Lucas service centre ought not to have any difficulty identifying your old one; if you buy a new regulator it will be the improved type. If you're thinking of cannibalising another alternator for a regulator, watch the colour of the leads. Some ACR regulators were made for temperature sensing and the main lead to the starter solenoid went via a heat sensor mounted as close to the battery as possible. This lowered the voltage when the battery was hot and raised it when the battery was cold. The connecting wires on a normal machine sensing regulator are coloured black, yellow, red and white. On a temperature sensing regulator the white is replaced by an orange lead.

If you want, you can check and change the brushes without taking anything else off. Each brush is attached to a small metal strip held to the top of the brush housing by two tiny hexagon-headed screws. However, it's easier, and better, to take the brush housing off so that you can clean the slip ring and also clean the inside of the brush housing which probably will be covered in carbon

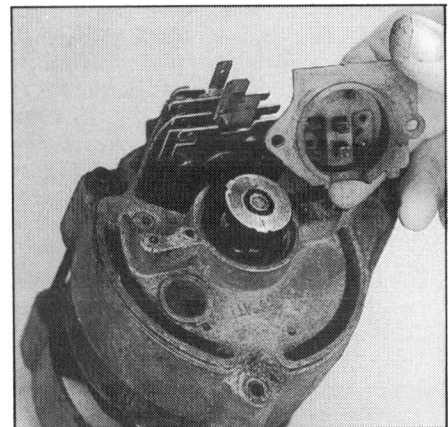

The brushes are in a white nylon housing which probably will be coated inside with carbon dust.

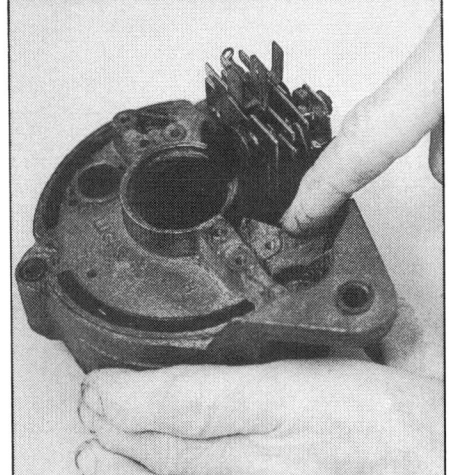

On an ACR alternator built before about 1980 you may find a surge protection diode bolted to this hole.

Before you can take out the rectifier pack you need to unsolder three wires. Use a pair of pliers as a heat sink to avoid damage to the new pack when you solder it in.

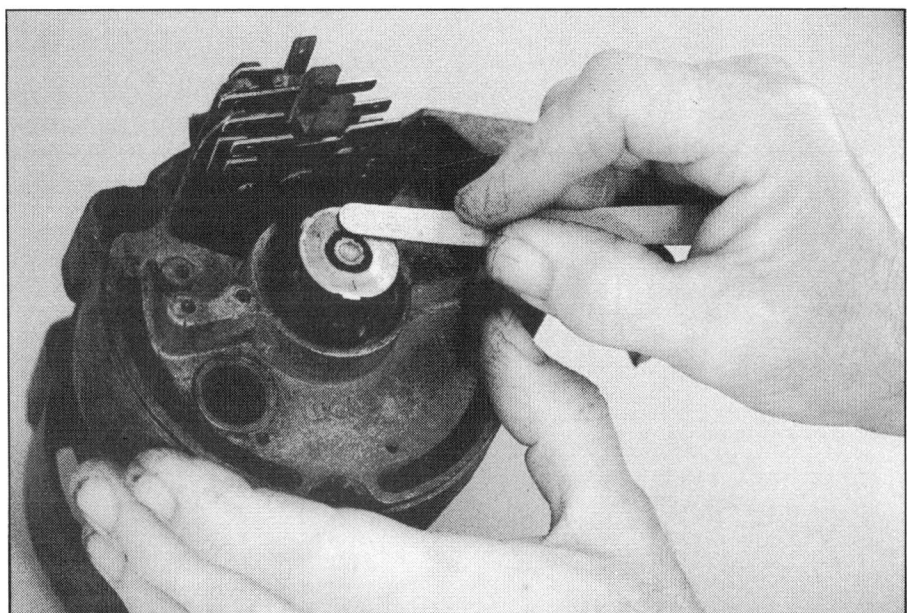

Clean up the slip ring with sandaper. A card 'nail file' is ideal.

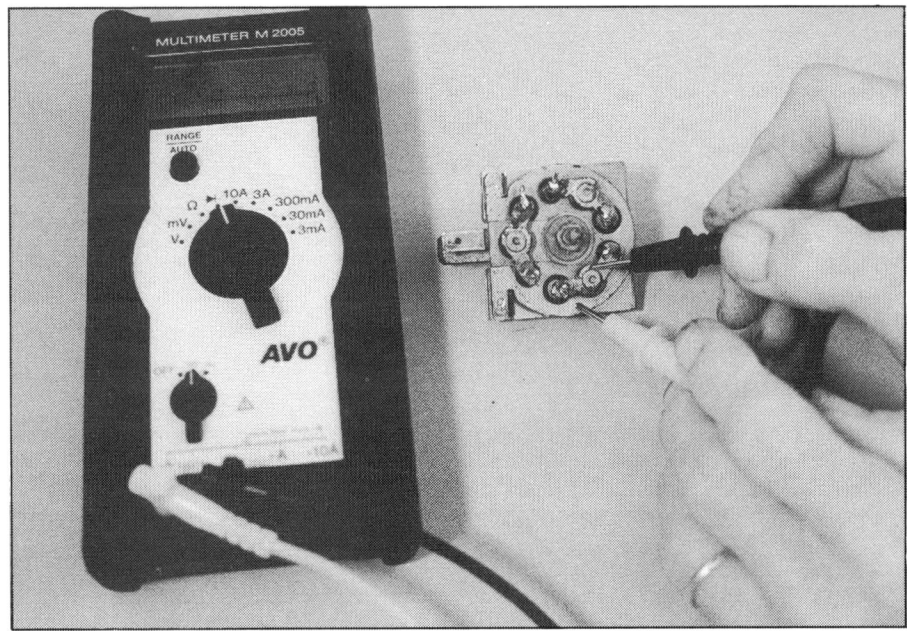

The only way to test diodes is with a meter or a lamp and battery. They should pass current in one direction but not the other.

dust which can short things out. The regulator is attached to the top of the brush housing, so you have to disconnect it first. Make a note of where the connections go. Before you start taking anything off, there's a simple, though not infallible, check you can carry out on the regulator. Take the alternator back over to the car, plug it in and switch the ignition on. Earth the body of the alternator and then earth the regulator box body to the alternator body. With the regulator body earthed, the ignition warning light should not come on. If it does, the regulator is possibly faulty.

Lucas say that if the brushes are less than $\frac{5}{16}$in long they should be renewed. Unlike brushes on a dynamo, brushes on an alternator don't have to be bedded in as they run on a flat slip ring and not on a curved commutator. The outer brush wears quicker than the central one but, for some reason, the central brush seems to wear the slip ring faster. On a well-worn alternator you sometimes find it's almost drilled a hole in the centre connecter on the slip ring. If the slip ring is in good condition apart from being glazed you can clean it up with fine sandpaper, not emery as this leaves carborundum dust embedded in it. If it's worn, it needs renewing, but we'll come to that in a moment.

If someone (not you, of course!) tried to connect the battery the wrong way round, or put jump leads on the wrong way round, or even did some electric welding on the body without disconnecting the alternator first, there's a chance that they've blown the diodes in the rectifier. Poor connection be-

tween the brushes and the slip ring, particularly intermittent connection, can also blow diodes. A warning of this is an ignition warning light that flickers on and off when you rev up. If this happens to you, check the brushes and the slip ring as soon as possible.

There are nine diodes and you can't replace them individually; you have to buy a complete new rectifier pack. To test them, you have to take out the rectifier, and this means unsoldering three wires. They're all in yellow sleeving, so note where each one goes before you take them off. Heat is another enemy of diodes, so use a heat sink on the diode stem to divert the heat from your soldering iron. A pair of pointed-nosed pliers is ideal. Once it's unsoldered, the rectifier comes off after undoing one nut.

Some people do a rough check on diodes by feeling them and say that, if they're loose in their housing, they're dud but this is a very unreliable guide. The only way to check them properly is with a meter, or with a 12V battery and a 1.5A bulb such as a dash lamp bulb. Diodes are electrical one-way valves in that they pass current one way and

Check-out for an ACR

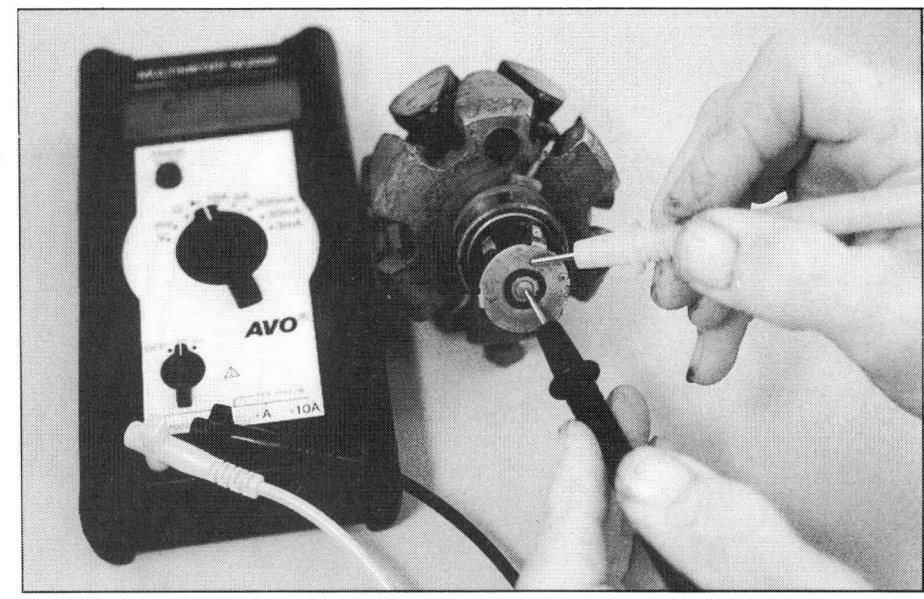

You can also use your meter or lamp and battery to test for continuity through the slip ring.

not the other, which is how they rectify the AC current generated by the alternator into the DC current your battery needs. If any of the diodes passes current both ways, the whole rectifier needs replacing.

If the slip ring is badly worn, the whole plastic assembly can be replaced quite easily. It's possible, but tricky, to replace this without any further dismantling but it makes things easier to get at if you undo the two long through-bolts that hold the alternator body together and tap the back part of the body off its bearing. You have to unsolder two wires which go to the slip ring, one to the outer track and one to the centre, and once again they're both yellow-sleeved so make a note of which goes where. After that the slip ring just pulls off the end of the rotor shaft.

The rear bearing of an alternator seldom gives trouble but the front one, at the pulley end, sometimes does. To get at this you take the pulley off and knock the Woodruff key out of the shaft. The bearing is captive in the alternator body and usually very tight on the rotor shaft. If you decide to knock the shaft through, put the nut back on to get it started or you'll damage the threads for certain. Even the last part, when the nut won't pass through the bearing, is tight and it's far better to use a press or a puller to get the housing off.

The bearing is held in the housing by a circlip but there aren't any ears on this to take circlip pliers; you lever it up out of its groove with a small screwdriver. It makes it easier to take the old bearing out and put the new one in if you first heat the body slightly to expand it. Boiling water is usually hot enough.

Reassembly is, to use a well-worn phrase, the reverse of taking things apart but, before you finally bolt the back cover on, give the three terminals on the rectifier pack a good clean with sandpaper. It also pays to check the inside of the plug connecter which fits on them and fit a new one if the contacts are corroded. You'll probably notice that the two larger terminals on the rectifier pack are joined together. Only one of them is used on the 15ACR, 16ACR and 17ACR alternators but both are used on the heavier-duty 18ACR.

If you've got a lot of electrical equipment on the car you can change up from a 15ACR or 16ACR to an 17ACR, relative outputs of the 15, 16 and 17 being 28, 34 and 36A. Most popular is the 17ACR because it has a low cutting-in speed of 950rpm compared with the 1,000 or 1,250rpm of the others, and will charge on a fastish tick-over, quite an advantage in stop-go traffic when you've got a lot of electrics on. □

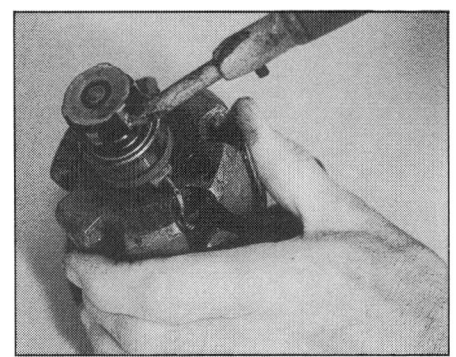

To get the slip ring assembly off there are two more wires to unsolder.

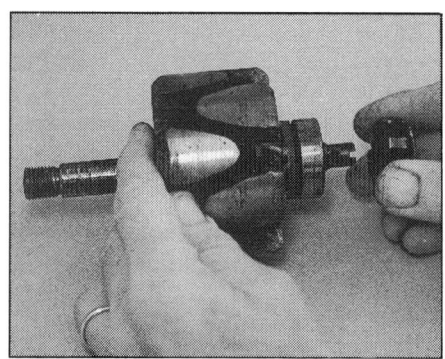

After unsoldering the wires, the slip ring just pulls off the end of the rotor shaft.

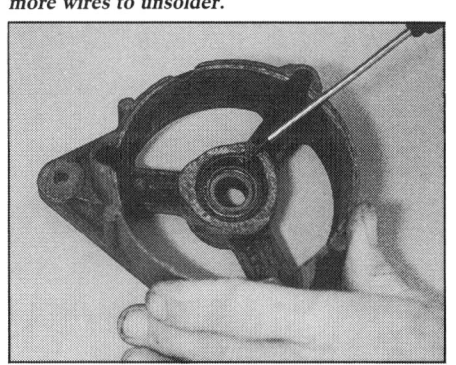

The drive-end bearing is held in the body by a circlip which levers out with a small screwdriver.

Don't forget to clean the Lucar terminals on the rectifier pack and also check the inside of the plug-in block.

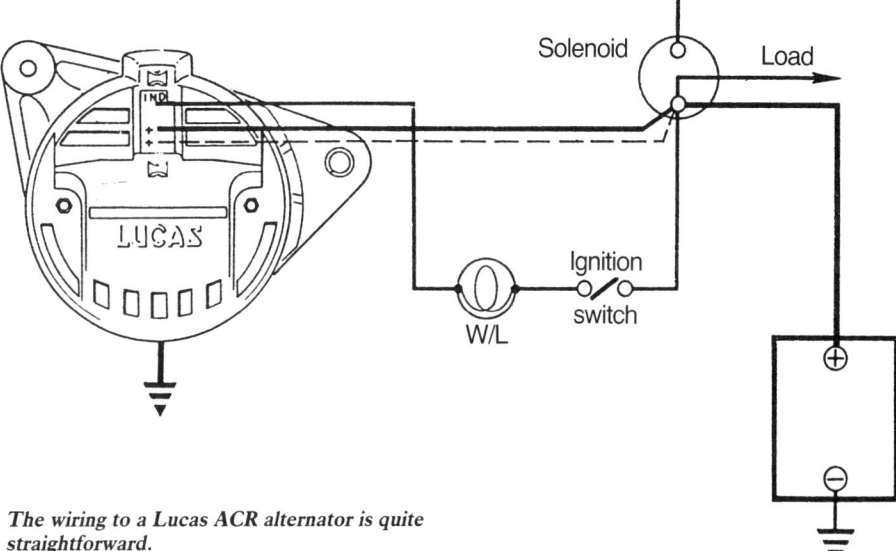

The wiring to a Lucas ACR alternator is quite straightforward.

INSTRUMENTS

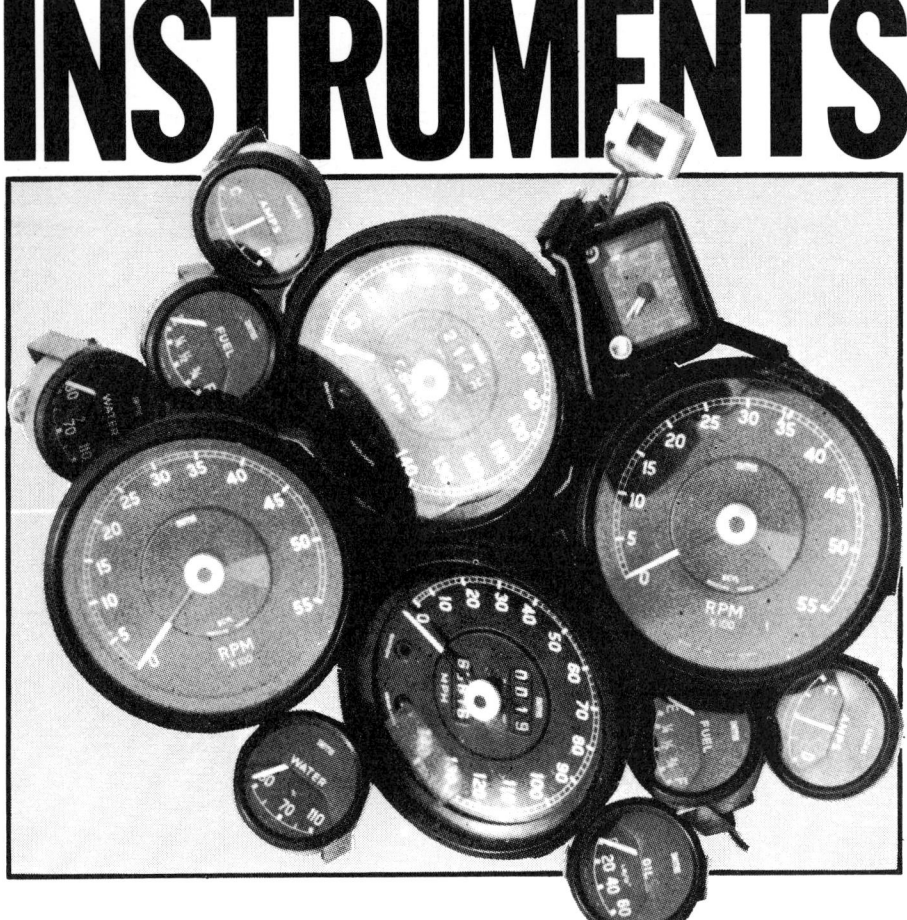

John Williams discovers that there is more to dashboard instruments than meets the eye.

Although the information which is provided by the dashboard instruments is probably heeded by the majority of drivers, the instruments themselves and the way in which they work tends to be ignored, at least until something goes wrong. In this article I propose to describe how some of the more common instruments work, and what you can do if they cease to function.

Speedometers

These are normally driven by a flexible cable from the gearbox which rotates a magnet situated within a metal cup in the instrument. The cup, which is linked to the pointer in the speedometer, tends to rotate due to the magnetic forces, but is restrained by a hairspring to an extent which is determined by the calibration of the instrument.

Most speedometer faults can be traced to the condition of the flexible drive cable or its connections, and these should be checked first if the instrument fails to operate. Assuming that the correct speedometer is fitted, inaccuracy may be caused by non-standard tyres or rear axle, or by tyres which are badly worn or inflated to incorrect pressures. If the pointer of the instrument waves over a wide arc or works sluggishly, the cause may be excessive oil in the speedometer. A kinked or crushed flexible drive will cause the pointer to waver continuously through an angle of about twenty degrees, and intermittent wavering indicates that the flexible drive is not engaging

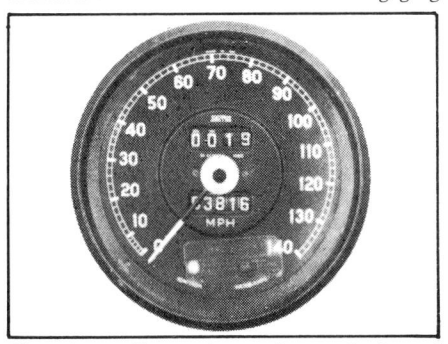

There is usually a code number on the face of a speedometer which enables you to check with the manufacturers that it is the correct instrument for the car.

fully. A breakdown of the mechanism which drives the odometer/trip counter due to stripped gears etc. must be referred to a specialist as spare parts are not otherwise available. Noises also assist in pinpointing the sources of faults. The speedometer will require specialist repairs or even replacement if it emits screeching noises, or if the odometer/trip counter mechanism produces a ticking noise which varies with the speed of the vehicle. Tapping noises can usually be traced to a sharp bend in the flexible drive, and either end of the flexible drive can produce other sounds associated with wear in the ends of the inner cable.

Maintenance of a speedometer is limited to very sparing applications of instrument oil at 10,000 miles intervals to the bearing where the flexible drive enters the instrument.

The only maintenance which can be carried out by the D.I.Y. enthusiast amounts to a very sparing lubrication of the bearing where the cable enters the rear of the instrument (preferably using very light instrument oil), and equally sparing lubrication of the cable itself (using grease) at about 10,000 mile intervals. The cable should be routed smoothly from the instrument to the gearbox, avoiding sharp bends, and it should not be crushed by any clamps through which it passes. Repairs to the speedometer have to be done by a specialist who has access to a supply of spare parts and who is equipped to re-calibrate the instrument after reassembly.

Ammeters

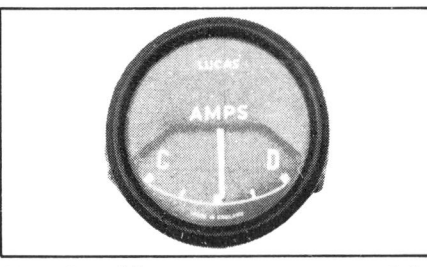

Note that different ammeters are used for dynamo and alternator systems.

The ammeter measures the actual current flow to or from the battery. It consists of a heavy coil of one or two loops of wire around a small magnet which is linked to the needle. Current flowing through the wire creates a magnetic field which interacts with the magnet and moves the needle one way or the other depending on the direction of the current. The larger the current the larger the deflection of the needle.

An ammeter gives an instantaneous indication of the state of the battery. If, for

There are a number of methods of connecting an ammeter and it must be connected correctly to avoid ruining the instrument and damaging other parts of the electrical system. The correct wire should also be used and this is 44/0.30 for dynamo systems or 65/0.30 where an alternator is fitted.

example, the load on the battery is suddenly increased by turning on headlights and foglights, the increased current flow from the battery will be shown immediately by the ammeter.

It is not an ideal arrangement to have an ammeter and an alternator in the same electrical system. The efficiency of the system depends heavily on good soldered or screwed connections throughout. With an ammeter in the system the voltage output of an alternator tends to be greater due to increased resistance. This does not matter so much with a dynamo, but it can damage the transistors in the alternator. This is why many modern cars have a battery condition gauge rather than an ammeter — and the ammeter installation is more expensive.

There is little that can go wrong with an ammeter. As current passes through the instrument to all systems in the car, it follows that if, for example, the headlights are working then the ammeter must be receiving current. A suspect reading may be due to bad connections, in which case there will be a build-up of heat at the terminals. Otherwise there may be some mechanical defect such as a tendency for an internal linkage to stick, or worn bearings. There is usually nothing that the owner can do about mechanical problems. If you are buying an ammeter and want to check it's mechanical condition, shake the instrument gently and the needle should swing freely from side to side.

A disadvantage of the ammeter is that all the current used in the vehicle has to pass through the ammeter circuit which therefore has to employ heavy wiring, and this presents particular problems in relation to rear-engined cars.

Battery Condition Gauge

This is in effect a voltmeter connected directly to the battery and it is intended to reflect the actual battery voltage. It works by means of a heating coil through which the current passes and a bimetallic strip which is connected to the needle and is affected by the heat from the coil. A battery condition indicator is a poor substitute for an ammeter as it does not give a true indication of what is happening to the battery at any given time. For example, if a short circuit occured suddenly in the electrical system it would take at least several minutes for the battery voltage to decrease and for this

decrease to be shown by the battery condition indicator, whereas the ammeter would show the increased rate of discharge from the battery immediately.

Rev Counters (or Tachometers)

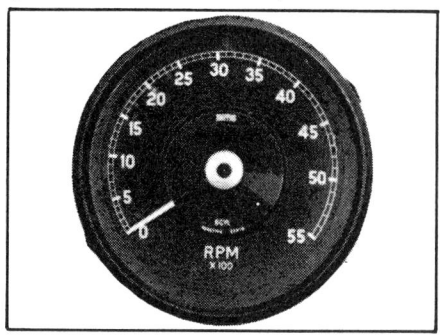

Tachometers are not necessarily interchangeable between different cars but universal types are available.

There are mechanical tachometers which are cable driven from the camshaft and work in the same way as the speedometer. There are also various types of electrical tachometers. An early type used on Jaguars was simply a voltmeter which received current from an AC generator driven by the camshaft. The output of an AC generator is directly proportional to the speed at which it is driven, and the instrument is calibrated in RPM rather than volts.

The electronic current impulse tachometer is a later type, connected to the ignition circuit between the coil and distributor and operated by the current impulses which occur as the contact points open and close. The impulses cause a magnetic field which moves the pointer around the scale of the instrument. This type is usually polarised, that is it has to be matched to the polarity of the battery connections, but there are universal types which can be wired to fit positive earth or negative earth cars.

Then there are voltage triggered tachometers which are in effect transistorised voltmeters measuring the frequency of the voltage drop across the ignition coil as the contact points open and close. Both these and the previous types are transistorised and must be taken to specialists for repairs. Fault finding is limited to checking that the wiring is in order and the connections clean and tight.

Fuel, Temperature and Oil Pressure Gauges

These instruments may be considered as a group because in any one car they usually work in the same way. There are three types of these gauges, dynamometer, hot wire, and mechanical. The dynamometer type is an earlier design. It consists of a magnet which is connected to the needle and located between two coils. Current from the battery passes through the control coil to earth. The operating coil also receives current from the battery but this then passes through the variable resistance in the transmitter or sender unit before passing to earth. The magnetic

field resulting from current passing through these coils moves the magnet and the needle. Therefore when, for example, the fuel tank is empty, there is minimal resistance and therefore maximum current through the operating winding which results in the needle being pulled over to the 'empty' position on the gauge. The temperature and oil pressure transmitters also vary the current passing through the operating coil but these transmitters are sealed and have to be replaced if they fail. The fuel tank sender unit can at least be checked to ensure that the electrical connections are clean and tight, the arm free to pivot with the level of fuel, and that the float itself is not punctured. Variation in battery voltage will affect both coils equally and do not therefore interfere with the reading shown on this type of gauge. The reading given is instantaneous and in the case of the fuel gauge it is affected by movement of fuel in the tank.

More often than not a fuel gaguge fault can be traced to the tank unit where the wire has become loose or damaged or the unit is not earthing adequately.

The hot wire or thermal type of gauge also contains a wire through which current passes to the transmitter unit and then to earth. The wire, which is linked to the needle is heated by the current causing it to expand and contact with the variation in temperature. A similar type of gauge employs a bimetallic strip heated by a coil. This type of gauge is designed to work at below battery voltage (usually between 8 volts and 10 volts) and all the instruments of this type in the car will receive current via an instrument voltage stabiliser, so that they are not affected by variations in battery voltage. Therefore if all these instruments are showing a reading which is incorrect it is likely that the instrument voltage stabiliser is at fault. Readings shown on this type of instrument are subject to a delay whilst the coil reaches the appropriate temperature and the needle creeps across the dial.

There are also mechanical pressure and temperature gauges, and some cars of the classic period had combined oil pressure and water temperature gauges. In the case of the temperature gauge a bulb containing an alcohol based liquid is situated in one of the water passages in the engine and it is connected to the gauge by a very thin tube. Very great care is needed whenever it is necessary to disturb the capillary tube, for example when disconnecting the assembly from the engine. These capillary tubes become brittle with age and when they are disturbed there is a high risk that they will split and the

CONTINUED ON PAGE 37

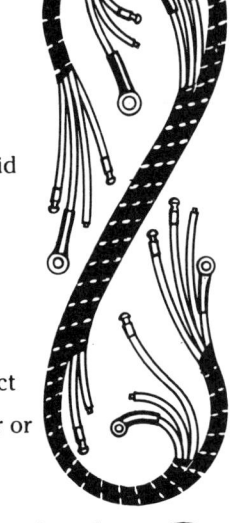

Trafficators

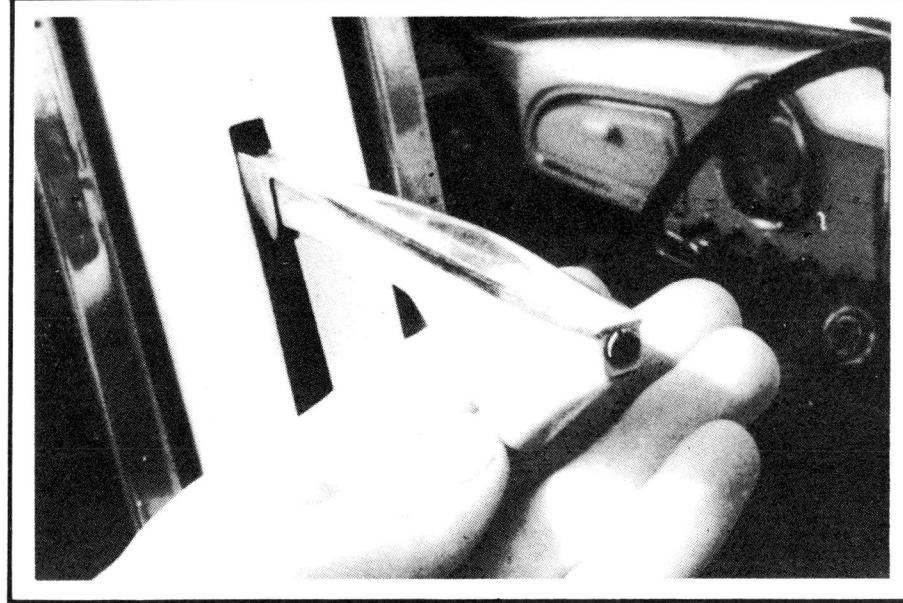

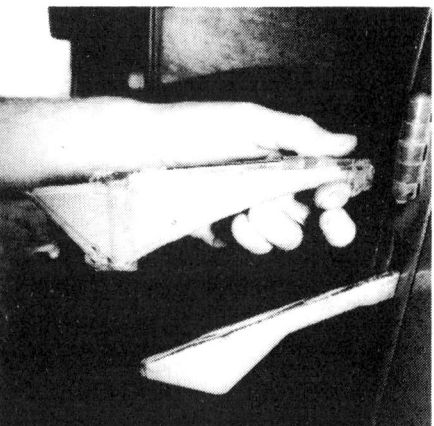

The very earliest type of Trafficator (the SF26) had the very 'twenties' looking arm as held here; later types were less ornate and are considerably easier to find.

How they work and what to do if they don't. We investigate the subject with the help of the Vintage & Classic Car Spares Co.

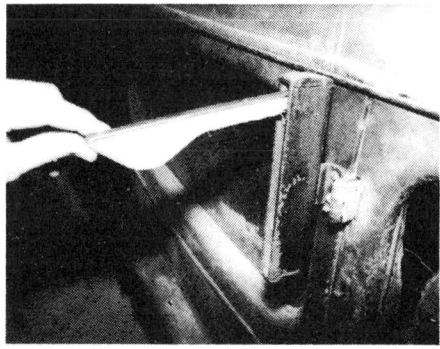

Some Trafficators, both pre- and post-war, were contained in an exterior die-cast or pressed steel case; the actual 'works' were identical to the concealed type however.

Semaphore indicators lingered on into the sixties (the Morris Minor being one of the last cars to succumb to 'flashers') and while it's a bit dodgy to rely on other road users noticing that funny little lit-up stick, if you have them on your car then at least you should make sure that they work as effectively as possible. It isn't difficult, and although there are ten or more different types, they all work on exactly the same principle.

The most common type of Trafficator (a Lucas trade name, though other firms made them too) in use today is the type SF80 which was introduced around 1950 and was standardised about 1956. The SF80 differs from its forerunners through the electrical contact for the bulb being made at the end of the arm itself, rather than by a separate wire running into the arm. Other than that, the mode of operation is the same — when the indicator is switched on, an electrical current energises a solenoid, the armature of which is drawn into the core and thus lifts the arm; additionally, a current is passed to the bulb which lights up.

Above:
The most common type of semaphore indicator, the SF80, fitted here to a Morris Minor 1000.

When switched off, the arm returns by gravity — or it should do!

Largely because it *is* so simple, the electrical part of a Trafficator rarely causes any trouble; about the worst thing that can happen is that the coil burns out, in which case the cheapest way out is to look around for another unit, new or second-hand, although it is possible to have the solenoid re-wound, at some expense. Or on the pre-SF80 types, the wire which carries the current to the bulb can fray and earth somewhere — easily repaired, a model shop being a good place to find a suitable length of sleeving. Other than these, the only further electrical item to check is the auxilliary ignition fuse in the car's control box, which supplies the trafficator circuit when the ignition is switched on. Ensure that the fuse is making proper contact and isn't burnt-out. The indicator switch and associated wiring can be checked if in doubt through 'bypassing' with a separate lead connected to the Trafficator.

However, if there is still no joy the unit may have to be removed from the car. Sometimes this is done by removing an outer escutcheon plate, or otherwise the trim on the door pillar (or whatever) must be taken off if access to the couple of set-screws which hold the unit in place, is from inside the car. Only one wire usually joins the indicator to the car, normally

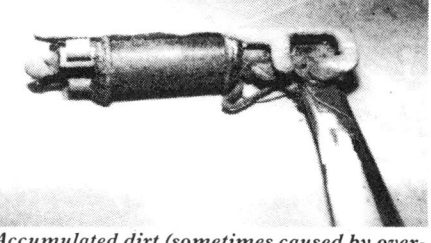

Accumulated dirt (sometimes caused by over-oiling) can easily prevent the indicator from working properly. A thorough clean, whether you intend to dismantle or not is the best procedure.

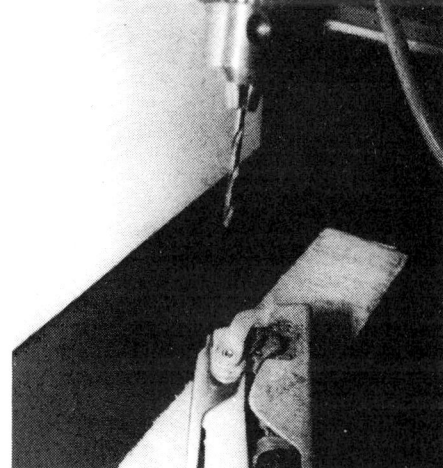

To remove the arm from the base, the end of the securing rivet must be carefully drilled to allow it to be pushed through with a screwdriver.

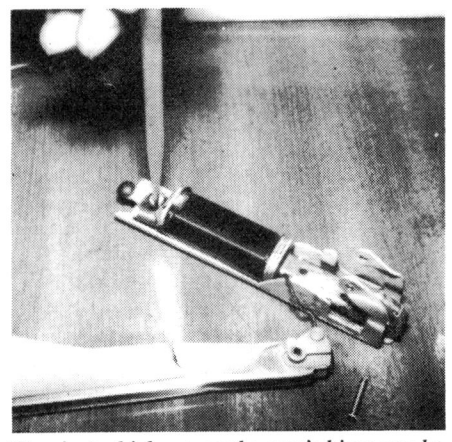

The rivet which acts as the arm's hinge can be seen bottom right; to remove solenoid assembly from casing means undoing this small screw.

Make sure you know how the insulators are positioned under the screw which locates the solenoid and connector.

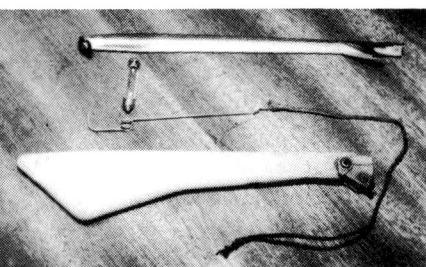

The arm assembly in dismantled state; wire lead can sometimes fray and short out. Circuit to bulb is completed by bulb coming into contact with metal sheath as arm rises.

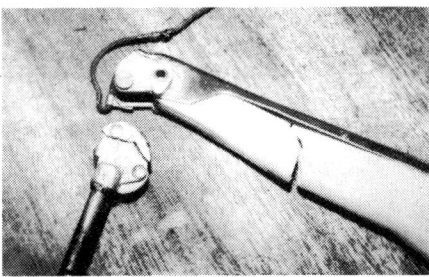

This is the armature which pushes the arm up. A casting, the end assembly is prone to breakage as it weakens with age. Cracked plastic lenses are not that uncommon either — amber types are getting quite hard to find.

Note that solenoid sits on a paper insulator; some of the wiring connections are by solder.

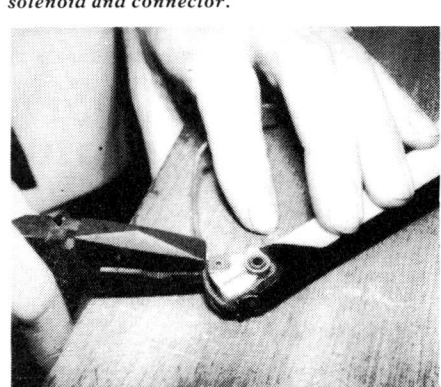

On pre-SF80 units as illustrated here, wire to arm is released by carefully bending back clip.

This is an SF80 unit, minus lens assembly. Note absence of separate lead to bulb — remainder of unit is the same in principle as the older types.

OPERATING TESTS

For those of you with the necessary equipment, this is the official Lucas test procedure as laid down when the Trafficator was current.

All MODELS Except SF80

1. With the arm bulb in the circuit, the unit must operate satisfactorily, i.e., unlock and lift arm to the horizontal position when the unit is inclined backwards at 5° to the vertical at a voltage between the terminals of:

   4.5 volts for 6 volt units.
   9.0 volts for 12 volt units.

   (Units to be cold when tested).

2. The arm must fall right home and lock when released from a position making an angle of 60° to the vertical.

3. The unit must be operated intermittently for five minutes with periods of 10 seconds on and 10 seconds off, with the nominal voltage across its terminals, and mounted at 5° to the vertical (as stated in Test 1). The unit must function correctly throughout the test.

MODELS SF80

1. With the arm bulb in circuit, the unit must operate satisfactorily, i.e., unlock and lift arm to bring the lower edge of the arm to cover to within 2° of the horizontal when the unit is inclined backwards at 5° to the vertical with a voltage between the terminals of:

   4.5 volts for 6 volt units.
   9.0 volts for 12 volt units.

   (Units to be cold when tested).

2. The arm must fall right home and lock when released from a position making an angle of 45° to the vertical.

3. The unit must be operated intermittently for five minutes, with periods of 10 seconds on and 10 seconds off, with a nominal voltage across its terminals. The unit must function correctly throughout the test.

BOBBIN RESISTANCES

	6 volt	12 volt
All models except SF80	0.82-0·93 ohms	3.9-4.4 ohms
Model SF80	1.5-1.7 ohms	4.5-5.2 ohms

via a push-in connector, though a further earth wire may be present especially on an older wood-framed car. You can now take the assembly to the work-bench for a proper examination.

The easiest thing to do is remove the small screw from the end of the arm and slide the metal casing off so that you can take a look at the bulb, but to actually detach the arm from the unit means drilling the end off the rivet on which it hinges. Then to dismantle the solenoid and get at the various connections and the base-plate, the screw just above the arm's rubber buffer must be undone (if the rubber is still there — and if it is, check that it hasn't gone sticky with age and glues the arm closed). Be careful to note the various insulating pieces disclosed as you take the assembly apart, particularly the order they go back in. Solenoid and armature can now be cleaned using paraffin or contact cleaner. Examine the alloy 'carrier' with its spring loaded jaws at the end of the armature — this is subject to fracture as it deteriorates with age, and may be the cause of the indicator not working.

After cleaning, reassembly can begin providing you've traced the fault (if any). Do not over-lubricate as this will attract dirt, but a drop of oil should be applied to the arm's

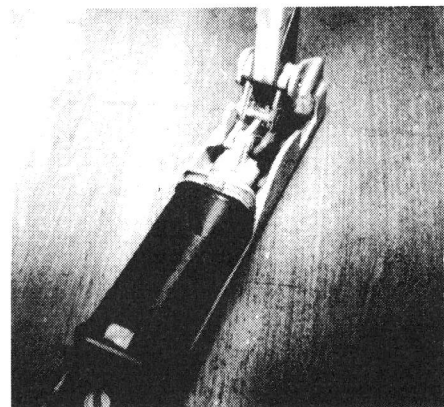

Continued

pivot, and petroleum jelly or thin grease to the spring-loaded jaws. Points to watch at this stage are that the armature moves easily inside the solenoid, that the various insulators are in their proper positions, and that the spring on which the festoon bulb sits goes back into the lens first, and in the first slot. When rivetting the arm back on, ensure that the 'jaws' on the armature are in position on the indicator arm base first — easy to forget.

On reassembly, ensure that the jaws of the operating lever (armature) are over the indicator arm properly before rivetting the latter back into place (simply put rivet through arm and mounting, lay unit on its side on hard surface, and use something like a centre punch to open out the rivet).

The chances are you won't need any spare parts as just a build-up of dirt or a poor connection somewhere is often the culprit when a Trafficator is reluctant to function correctly. However, for some reason (perhaps because they are the most-used) it is the later (usually amber) plastic lenses which are hard to get, the earlier yellow-type being more easily found. Incidentally, many older cars have collected the later SF80 type unit as a result of a previous owner updating his car by fitting a new unit rather than repairing the old one. At anyrate, the semaphore indicator is easier to strip and rebuild than an article of this length might appear to suggest, so long as you note how the unit comes apart and don't lose any of the bits! □

CONTINUED FROM PAGE 33

INSTRUMENTS

On a well designed dashboard the instruments are designed so that as many needles as possible will be in the vertical position when showing the 'normal' reading, including the tachometer whose needle will be vertical at maximum permissible revolutions and will run parallel to the speedometer needle.

vital liquid will be lost. If any part of this assembly becomes damaged, the gauge, bulb and tube have to be replaced as a unit, or repaired by a specialist (see 'Going Spare' in this issue). When the liquid in the bulb is heated in the engine the liquid expands and the resulting pressure is transmitted along the capillary tube to the gauge which then works in the same way as the oil pressure gauge described below.

The majority of mechanical pressure gauges employ a Bourdon tube which receives oil by a pipe from the engine. The Bourdon tube is flattened in cross section and curled lengthwise. It is sealed at one end and fixed to the instrument, and the other end is linked to the needle on the gauge. Oil pressure tends to uncurl the Bourdon tube (rather like blowing into a party whistle) thus moving the needle.

Repairs to instruments

Bearing in mind that most instruments used in motor cars are not likely to be totally accurate, and that drivers tend to recognise an abnormal reading rather than worry about the exact information given by a normal reading, there is little point in attempting to improve the performance of instruments. Unfortunately there is also little that the D.I.Y. enthusiast can do about repairing a faulty instrument. I

have already referred to the spares and re-calibration problems associated with speedometers and mechanical tachometers, and the fact that spares — when they are available at all — are only available to specialist repairers, will limit the scope for home repairs on other instruments too.

However, it should be possible for the home repairer to ensure that mechanical linkages within instruments are in working order and that electrical connections are good. It may even be possible to re-solder wires which have come adrift, given a sufficiently small soldering iron and a delicate touch. It should certainly be possible for the average owner to check that the fault is in the instrument and not in fact due to a poor connection or broken wire elsewhere in the circuit. All that is needed for circuit testing being a bulb holder fitted with (preferably) a 21 watt bulb and two wires one of which should have a crocodile clip. A wiring diagram showing the colours of the wiring in the circuit to be tested is almost essential. Simple circuit testing then consists of earthing various points on the circuit via the test light, starting at or near the source of supply, until a point is reached where the light fails to work. The fault in the circuit will be between this point and the last point at which the light did work. □

The writer would like to thank Mr Bob Krafft for his assistance in the preparation of this article.

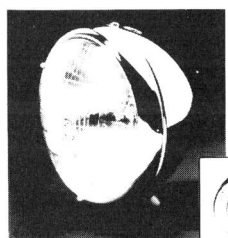

Overhauling Horns

The term 'overhaul' is perhaps a bit of an exaggeration, for horns are usually a very simple mechanism and in the usual course of events not a lot goes wrong with them. Later types are not designed to be dismantled in any case and, if they give trouble the recommendation is to throw them away and fit new ones rather than try to repair them.

On earlier models, however, it was possible to take them apart and do a certain amount of work. Two types are involved — the 'high frequency' or HF horn and the Windtone. The method of operation of both is basically similar. The high frequency horn uses an electro-magnet and contact breaker and the Windtone a solenoid and contact breaker to vibrate a diaphragm.

In the HF type a baffle in front of the diaphragm sets up vibrations in the air and this produces the sound. In the Windtone, there is a column of air inside the trumpet (either straight or coiled) and this is set in motion by the diaphragm to produce the sound — a more musical note than that of the HF horn.

Very often Windtones are fitted in pairs. They sound simultaneously, one with a high note and one with a low. They are marked H and L and matched to produce a harmonious sound.

How they are dismantled and adjusted can be seen from the photographs and some of the main steps are described in the captions.

HF Horns

Disconnect the battery before taking the horns off the car. The rather elaborate chromium-cased horns we used came in fact from an old Standard but they were also used by Bentley. All the chromework, however, is merely decoration and the horns themselves are no different, except in size, from those used on a lot of other cars of the same period.

Overhaul consists mainly of cleaning up the inside and then using thin emery cloth to take any carbon off the points. Check all the wires to make sure they are securely fixed and the insulation is intact. Look also at the point where the exterior wiring joins the horn. New thimbles can be obtained from a Lucas agent if required.

Adjustment is the final job and to do it, the horn will have to be connected up to a battery. Start by turning the adjuster anti-clockwise until the horn just fails to sound when the battery connection is made. Then turn the adjuster a quarter of a turn clockwise and the horn should sound.

There is another possible adjustment to the HF horn, although this is not normally necessary or advised. It does in fact alter the note of the horn and is merely a matter of turning the adjuster a quarter turn one way and then the other, sounding the horn in between and then settling on the note you like best.

Joss Joselyn explains what makes horns honk

Some high frequency horns, like this one, are encased with a grille and a back cover, others look very similar to the horn itself, shown on the left, while yet a third version may have a chrome ring or disc. The back is held by a single central hollow screw and the grille by three small rim screws.

The wiring connections can corrode and cause problems. Clean out the inside of the holes and clean the thimbles or fit new ones which can be obtained from a Lucas agent. A keep plate holds them in place.

Mark across the rim to ensure correct rebuilding before taking out the six 0 BA screws around the rim, including two holding the mounting bracket. Take great care not to damage the gasket and save it because it's a spacer and a special thickness.

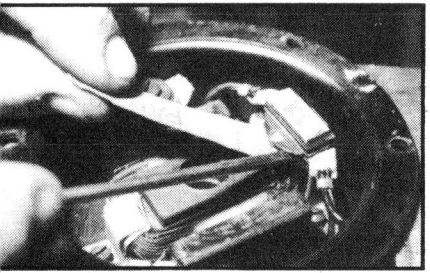

Prise the contacts apart gently and use a strip of fine emery cloth to clean the surface of both the top and bottom contacts.

When reassembling, line up the matching marks, which will ensure the armature is lined up across the pole shoes. The final job is adjustment shown here. You'll need to reconnect the battery first.

Windtones

Once again the horn we show in our photographs is a very top-drawer affair and in fact came off a Rolls Royce. The fact that the horn trumpet is long and straight does not mean it is any different mechanically from the more usual type with a coiled trumpet, although there is another slight difference with this one from most others in that there is an adjustment for altering the pitch of the sound.

With this particular unit, it is a matter of working the two adjustments together, by trial and error, until the best result is obtained.

Later Windtones only have the second adjustment which is concerned with operation and is intended to compensate for wear. It may be the top contact that is adjustable or it may be the bottom — probably the latter on coiled trumpet types.

The method of overhaul is much the same as for the HF type. All you can do is clean the

inside, check the wires are all attached and adjust, if necessary.

If you want to check that the horn itself is functioning without taking it out of the car, the best and simplest way is to run a wire straight from the battery (or perhaps two wires if you think the earth return could be suspect).

If the horn sounds when this check is made, check the mountings for good electrical contact, connections to the horns themselves and also the horn-push. These are often dependent on contact made by thin sprung metal components. A bit of corrosion or a loss of springiness can often result in no horns.

If the horn stops functioning because of contact wear, it is not generally necessary to take the horns out to do the adjusting. As shown in our final photograph in the HF horn section, there is an exterior adjustment. Later windtones also have the same facility and a simple half-turn on the screw might bring the horn back into noisy life. □

Not exactly a run of the mill windtone — in appearance anyway — this one came from a Rolls Royce. Mechanically, it is the same as any other windtone but instead of the extender trumpet, normally it is coiled round.

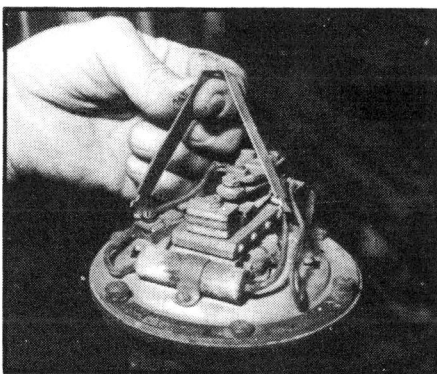

The dome is removed by taking out a single central screw and that can be followed by this two-legged mounting.

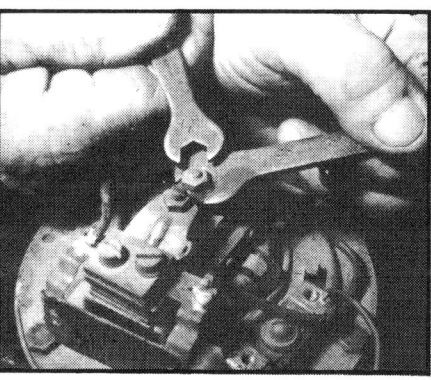

This is the adjustment for pitch — it's a matter of an adjusting nut and a locknut.

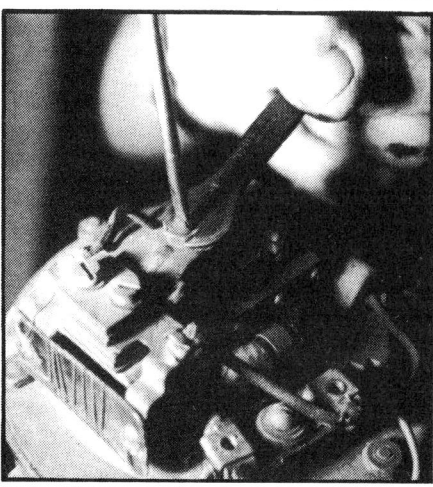

This adjustment controls the actual operation of the horn and is a matter of a screw and locknut.

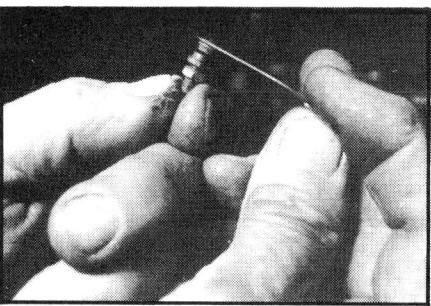

The type with the coiled horn tube often has a different adjustment, where it is the bottom contact that is adjusted up and down, as can be seen here.

Be careful if you're going to take the points apart. There's a lot of bits involved and the order is important. If you get it wrong, the horn won't work and rediscovering the correct permutation is not easy.

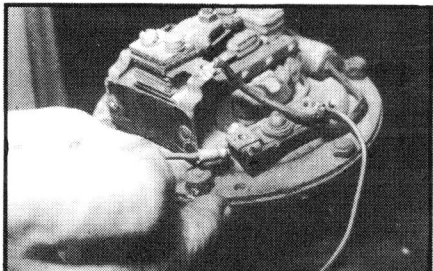

You can get the same sort of corrosion problems on these connections but fitting new bullet connectors is a simple soldering operation and totally standard.

Overhaul Your Heater

Improving the performance of the cars heating and ventilating system by John Williams.

A car's heating and ventilation system is usually left out of servicing schedules and is frequently ignored until something goes wrong, and since faults tend to occur when the system is most needed, now is a good time for a check-up to ensure that all will be well this Winter.

TYPES OF HEATER

The simplest type of heater is the recirculation heater whose sole function is to heat the air which is already inside the car. The Smiths Universal model is probably the best known heater of this type and it is usually fitted just above the front end of the transmission tunnel. The recirculation heater in its simplest form consists of a heat exchanging device which is piped to the engine cooling system and a fan controlled by a dashboard switch. This type of heater is usually fairly accessible but when removing it from the car (having first drained the cooling system) it is often easier to unbolt it from the bulkhead before attempting to disconnect the heater hoses.

More advanced versions of the recirculation heater incorporated other features such as a variable speed fan, further temperature control by means of a dashboard switch controlled water valve, and extra ducting and fittings to provide windscreen demisting.

This is a recirculation heater fitted with doors to control airflow and ducts leading to demisting vents.

The most common type of heater is the fresh air type to which air is ducted from outside the car, heated, and directed to the windscreen and/or the interior. In most fresh air heaters the fan is an integral part of the unit and the whole assembly is usually mounted against the bulkhead in the engine compartment. Some cars also have provision for 'face level' ventilation by unheated air which reaches the interior via dashboard vents. There are a number of variations in the design and layout of fresh air heaters and their controls, but all work in a similar way and most can be rendered inefficient or totally ineffective by the faults described below.

HEATER FAULTS

The first requirement of a heater is an adequate supply of water at a suitable temperature. Although there are variations between different engines the working temperature will be approximately 175F. Both temperature and water circulation depend upon the proper operation of the thermostat which together with its housing should be clean and unobstructed and which should open and close at the temperatures specified in the workshop manual. This can be checked by heating the thermostat and a thermometer in a pan of water. The hoses to and from the heater should be replaced if their condition is dubious, and two points are worth bearing in mind here. Firstly, a hose can collapse internally whilst appearing perfectly sound on the outside. Secondly, it requires only a small amount of internal deterioration

Fresh air heaters come in all shapes and sizes but at least part of the casing can usually be removed by releasing spring clips and/or undoing self-tapping screws When refitting a fresh air heater in a car, new bulkhead seals and grommets should be used.

On most fresh air systems the motor (complete with impeller or fan) can be fairly easily removed for servicing and on some cars the blower assembly is a separate unit as shown here.

A heater matrix is like a small radiator and can suffer from radiator problems such as leaks and blockages. Some replacements are still available and radiator specialists may be able to carry out reconditioning work depending on the extent of the damage.

which usually last a long time but can develop internal leaks which are not repairable.

Fresh air heaters can hardly be expected to perform well if there is any abstruction to the airflow. The first item to check is the air intake filter which has probably collected a few leaves or at least an assortment of dehydrated insects. Then examine the air ducting to ensure that it is clear of debris, undamaged, and free of leaks due to insecure or absent fasteners or just because it is too short or split. Remember that a heater which is allowed to draw air from the rear of an engine bay, could draw fumes into the interior of the car.

Other equipment which was not necessarily listed amongst the car manufacturers optional extras was available from accessory manufacturers. For example the Smiths rear screen demister was a blower unit designed to be built into the rear parcel shelf. Another was the Wingard fresh air heater — the model shown here is still available for the Austin A30 and A35.

in a hose accompanied by a build-up of silt and sludge to greatly reduce the capacity of that hose. Needless to say a hose should not be kinked around sharp corners or squeezed where it passes through a supporting bracket, bulkhead or other obstruction.

A check on the water circulation through the heater can be carried out by disconnecting the return pipe at the engine (but block the resulting hole at the engine) and extending it to reach into the top of the radiator. If the extension pipe incorporates or consists of a length of transparent tubing it will be possible not only to see the water flowing into the radiator with the engine running but also to detect air bubbles.

Whether a heater contains coiled water pipes or a heater matrix (like a radiator in miniature) its efficiency will be reduced by any

build-up of deposits. This problem may be reduced by flushing the heater with a hose pipe in both directions, but there is no complete cure short of having the matrix re-conditioned or obtaining another one.

Water leaks not only impair the performance of the heater, if only by creating an air lock, but also lead to far more serious problems of rust in the floor and chassis area if they remain undetected. Checks for leaks should be carried out on a regular basis and when new hoses are fitted the jubilee clips or other fasteners should be re-tightened after they have had time to become 'bedded-in'.

Water valves can suffer from leaks or blockages due to silting and corrosion and may need replacing. Earlier types of valves could be overhauled with a new diaphragm and could be adjusted.

The controls (cables, levers, linkages etc) should be checked to ensure that they are working freely, are not fouling adjacent components, and are correctly adjusted (where applicable) and adequately lubricated. On some cars the water valve and the scuttle (air intake) flap are operated by small servo units

Heater blowers can become jammed due to small objects falling (or being pushed by children) into the demister vents. Broken windscreen glass is a frequent cause of jamming and if the blower is turned on in this condition the motor may overload and burn out. It may be possible to obtain new brushes for the motor, other parts are not generally available and a replacement motor will be the only cure for more serious problems. Circuit testing is not difficult. Some cars are fitted with a two speed blower, this being achieved by passing the current through a resistance between the switch and the motor when turned on to 'slow', and direct from another switch terminal to the motor for 'fast'. On other cars the switch is combined with a variable resistance to enable the blower to be run at various speeds. Using a simple circuit tester comprising a bulb and bulbholder built into a length of wire with a probe on one end and a crocodile clip on the other, the crocodile clip is firmly attached to a suitable earthing point and the probe is used (with the circuit switched on) to intercept the circuit on each side of the components of the circuit, starting at the earth end of the circuit. The faulty component or circuit wire is the one on the earth side of the probe when the test light comes on. □

The writer wishes to thank Messrs R J. Grimes of Coulsdon for assistance in the preparation of this article.

Heaters on classics are not as efficient as the air-blending sort on modern cars; indeed some, like that on the Mk 2 Jaguar, have got a name for being little more than warmers. But many of them were far more efficient when they were new than they are after perhaps twenty years' service, and if you take a little trouble to overhaul yours you could be pleasantly surprised when the cold weather comes.

There are two main types of heater on classics, the recirculating type which just draws in and heats up the air inside the car, and the fresh air type which draws its air from outside. We'll deal first with the recirculating type, commonly nicknamed the fug-stirrer.

Quite a few of these were fitted after the car was sold, and though the makes vary they were all much of a muchness in design. They have a circular element, rather like a circular radiator core inside a cylindrical box. There are flaps on the box to control the flow — more or less — and a blower motor with a fan to draw the air in and push it out.

Usually the biggest trouble with them, as with the fresh air type, is that because the coolant flow is not very high, sludge and sediment collect in the bottom of the element and cut the flow down even more. Take the element out of its box, connect one of its pipes to the tap with a piece of hose and flush it through for about a quarter of an hour. You'll be surprised at the muck that comes out.

If the blower fan's been playing up have a look first at the switch. These are often a rheostat type with a wound wire resistor and a wiping blade that makes contact with it as you turn the knob. In the 'off' position it moves clear of the resistor. Probably the last quarter or so of the resistor track, which might be either the fastest or slowest speed depending on the design, seldom got used. In a lot of cases they've been run on fast speed all their life. The result is that except for one small part, the wire winding of the resistor is dirty and makes poor contact with the wiper. The switches usually have crimped-on covers, but they're not too difficult to take apart, and a few moments with fine emery paper will clean the winding and the wiper arm.

If that doesn't restore the motor speed, it probably means the motor itself needs attention, but I'll deal with the motors on both types of heater in a moment. Before leaving recirculating heaters I want to mention the shut-off tap and the hoses. The water shut-off tap is usually a simple screw-down tap on the cylinder head — shut it in summer and open it in winter — and it too gets clogged with silt. Fortunately most of them are easy enough to take apart and clean. On recirculating heaters, more so than on fresh air ones, the hoses often have sharp bends where they pass through the bulkhead. Many a heater has had its performance improved dramatically when the owner fitted new hoses with more gentle bends.

HOT TIME IN THE OLD CAR TONIGHT!

Well, fairly warm at least. Peter Wallage runs through overhauling your classic's heater for the coming winter.

A typical Smith's heater of the sort which sits on the bulkhead. Note the squirrel cage blower which is inefficient if the foam rubber sealing ring doesn't make a good seal.

On the other side of the bulkhead type heater are two long springs which toggle over-centre to hold the flaps for the demist and interior air flows. If the springs are weak or broken the flaps won't stay where they ought to be.

This is the element out of bulkhead heater. They often get full of sludge because of the low rate of coolant flow, and the fins get blocked with dust and rubbish.

The element from an internal 'fug-stirrer' heater is usually circular and a little more difficult to clean out.

Fresh air heaters are usually mounted on the bulkhead under the bonnet, and because they have to blow air through into the car rather than just stir up the air already inside, they have more powerful blowers often with a squirrel cage rotor instead of a fan. Some also have to draw the fresh air along trunking from the front of the car. Many a complaint of poor blowing power has been cured by attention to the rubber sealing ring where the outlet from the blower passes through the bulkhead, and many a complaint of fumes in the car has been cured by attention to the joints in the fresh air trunking.

Much the same applies to the internal flaps inside the heater casing which direct air either to the floor or to the demisters. These

This water control valve from a Smith's heater is not meant to be repaired, but with care it can be taken apart for cleaning if it's jammed.

This sort of blower fan, from an interior heater, has a split shaft and is held to the motor spindle by a screw collar.

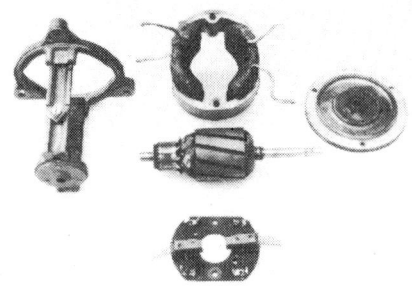

The stripped-out motor from a recirculating heater. You have to unsolder four wires to take it apart, so make a note of where they go.

On some two-speed heaters there's a large resistor like this. Check it for continuity before you decide the motor's burnt out.

are controlled by Bowden cables but held in the open or shut position by long coil springs which toggle over-centre. If these springs are weak or broken the flaps won't close tight and you won't be able to control the direction of the air flow properly — a common cause of bad demisting. The flaps seat on foam rubber or plastic, and you get the same troubles if these seals have gone hard and flat — as they usually have after all these years.

The water valve on this type of heater is usually controlled by a Bowden wire so it isn't a screw-down type, it's usually a diaphragm valve with a coarse cam to open and close it. Though the diaphragm doesn't usually give a lot of trouble, the cam often sticks or sludge stops the diaphragm from working.

They aren't meant to be repairable, but the two halves are held together by turned over lugs, and if you give the two halves of the valve a twist they will come apart. You may need plenty of Plus Gas and a vice, but they usually give in the end. If the worst comes to the worst you can make a good one up from several old ones. You might even have to, because they're a bit thin on the ground in most parts stores.

So far it's all been relatively plain sailing. Now we come to the awkward part, the motor itself. Heater motors are reasonably reliable, but they're quite cheap and cheerful and were never meant to be repaired so you can't get parts for them. In many cases you

can't get exchange motors any more, so you're forced to repair.

On some very cheap motors you might have to bend lugs to get the casing apart, or drill out rivets and replace them with nuts and bolts, but on most Smith's and Delaney heaters, which account for a large percentage of those fitted to British cars as original equipment, at least the motors come apart when you undo long through bolts.

Sometimes these long bolts also hold the motor to the heater casing, so be very careful when you take them out. On the recirculating type you seldom have trouble, but on the more powerful fresh air motors there are often spacers and coil springs threaded on the bolts to hold all the innards in position, and these fall out on the bench if you're not careful. If you don't get them back in the right order the motor will never work properly again.

A second point to watch is that the bearings are usually the self-aligning type which are spheres held in their housings by star-shaped spring clips. Sometimes these spheres seize on the shaft and turn in the housing, and when you pull the motor apart the star-shaped clip springs out. These clips are easy enough to put in with the proper adaptor on a hand press (which of course you haven't got) but can be really difficult to get back in with screwdrivers and levers, so be careful.

So far so good, but to get at the inside of the motor properly you've got to take the fan or squirrel cage rotor off the motor shaft. These are usually held by the collet of the fan being split and clamped with a screw-on hollow collar. In theory this is fine, but in practice the collet end seizes so tight to the shaft that you need heat and plenty of Plus Gas to get it off. In a couple of cases I've found them so tight I've given up and cannibalised another heater.

To get at the brush plate you have to unsolder wires, so make a note of where they fitted, but you won't usually find much wear on the carbon brushes themselves because the motor gets only intermittent use. Most troubles with brushes are because they stick in their housings, and to free them you have to bend back lugs on the end. Go gently because they break off very easily.

Clean the commutator in the usual way with very fine glasspaper. Don't use emery or carburundum paper because the grit on these is a conductor and will short out the gaps between the segments. Make sure the brushes are nice and free in their holders, and solder the wires back on. Try the motor on a 12 volt supply before you box the heater up, and while it's running give a couple of light taps on the end plates with a hammer to settle the self-aligning bearings in their correct alignment.

Few heater systems are self-purging so you'll probably have to bleed air out of one of the hose connectors when you put things back. Use plenty of rag to stop the anti-freeze from spoiling your nice paintwork under the bonnet or the carpets inside the car in the case of a fug-stirrer. □

The motor from a more powerful bulkhead heater. As well as unsoldering wires you have to watch out for distance pieces and springs when you take it apart.

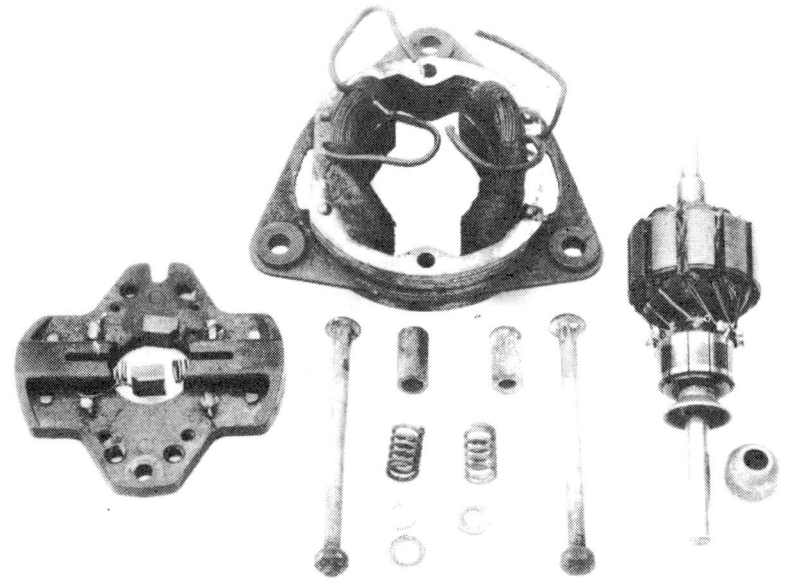

Valve Radio Repairs

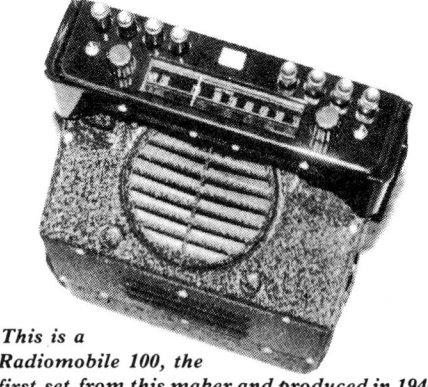

Car radio repairs — D.I.Y. or a job for the experts? John Williams investigates.

Wireless receiving equipment specially designed for installation in motor cars has been available since the early 1930s, but if like me you regard the whole subject of radio as one of those impenetrable mysteries you may well have concluded that there is nothing that you can do to improve the performance of your set much less repair it. I consulted Mr Rees of the Vintage Wireless Company at Bristol in order to find out what the do-it-yourself restorer could achieve (and should avoid) and what problems are likely to arise.

Prior to 1959 all car radios (or wireless sets as they were called then) were fitted with valves, from 1959 to about 1962 they contained a mixture of valves and transistors and more recent equipment has been fully transistorised. To accommodate valves and other bulky components earlier sets were large and in order that they could conveniently be fitted into a car the radio, amplifier, speakers and tuning device were often supplied as separate items. Early wireless sets were relatively expensive (the cost of a set just after the second World War amounted to about three times the average weekly wage at that time) but they were also very well built, they were repairable throughout, and Mr Rees claims that many valve radios which have been fully restored to as new condition will give a superior overall performance when compared with their modern counterparts.

The radio in the **Practical Classics** *Sunbeam Rapier is not working (yet) but I am going to check that all electrical and aerial connections are sound before I seek expert help. Note that all radio components should be securely mounted (to prevent vibration) and radios or amplifiers which contain valves should be in a well ventilated position.*

This is a Radiomobile 502T — note the separate amplifier.

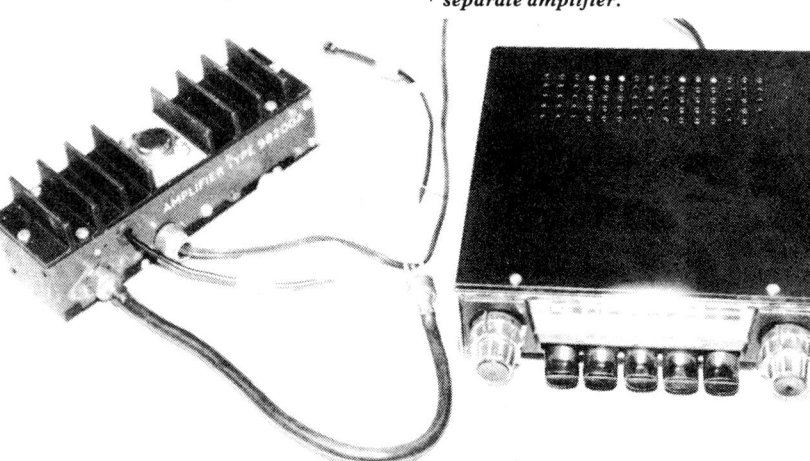

Mistakes to avoid

There are certain basic mistakes to avoid when dealing with a car radio. Firstly when connecting the set to the power supply do observe the correct polarity; this is vital when a radio incorporates transistors but admittedly it is not important on some earlier all valve sets. Never switch on a car radio installation unless the speaker is connected as this can cause a great deal of damage to the radio. Make sure that the correct fuse is fitted. Early wireless sets required fuses of up to 5 amps or even 8 amps in some instances, whereas a 1 amp fuse will suffice for most modern radios.

*This is a
Radiomobile 100, the
first set from this maker and produced in 1946.*

Installation and fault finding

If you buy a car radio of any type, but especially an older one it is advisable to ensure that it is complete. Internal components such as valves etc. can be replaced (a few valves are rather expensive but at least all are still available) but external parts such as knobs and the tuning dial itself may well be missing. These parts are often irreplaceable. If your set should have a separate amplifier, but it is missing, a specialist could probably make one for you if the correct amplifier, or a suitable alternative, was not available from stock.

Speakers are not a problem, in that you do not have to worry about polarity when wiring up the speaker unless more than one speaker is being installed, but the wiring connections to speakers should be soldered. Valve radios always had 3 ohm speakers but these are no longer available and if you need a modern substitute a 4 ohm speaker will be suitable. Bear in mind that a radio or amplifier which contains valves produces hear and should be installed in a well ventilated position. Cheap aerials are to be avoided. A good quality aerial will perform better and will also keep out water thus preventing internal short circuits between the aerial lead and its casing.

The average amateur restorer is unlikely to have the specialised knowledge or equipment which would enable him to test all aspects of his car radio or identify faults in internal components. No specialised knowledge is needed in order to check that earth connections are clean and tight though. The radio (and amplifier) casing is usually connected to a suitable earthing point on the car (which may be the actual mounting brackets), and although the casing of the aerial lead is usually intended to earth where the aerial plug enters

the radio, this is earthed on some early radios to the panel which carries the aerial.

Aerial Tuning

Aerial tuning is a fairly simple operation and the recommended method of achieving this is

If you have no special knowledge of radio equipment you will readily accept that repair work is a specialised job. Note that many early sets employed quite high voltages and can give a nasty shock to anyone who tampers inexpertly.

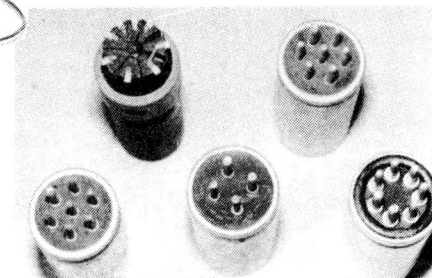

Some of the components which can be replaced include these vibrator-convertors . .

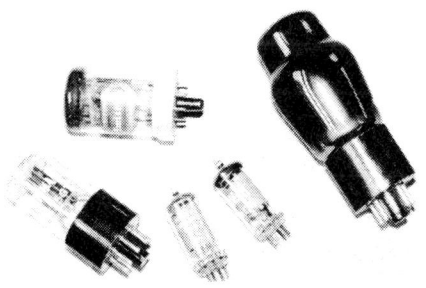

. . . and these valves. Although Britain once led the world in radio technology most of these are now made abroad.

as follows. Firstly select a distant station (preferably a foreign one) at around 220m on the medium waveband and turn the volume up. Then adjust the aerial trimmer (this is often a small screw which is accessible through a hole in the radio casing) to obtain the maximum available volume.

All electrical connections should be perfectly clean and a brass suede brush is a useful tool in this context. Electrolube is a special cleaner which is recommended by Mr Rees.

When a radio ceases to work the first thing to check is the fuse. On no account should a heavier fuse be installed but if fuses of the correct amperage are repeatedly blowing, expert advice should be obtained. It may seem an obvious statement but radio components deteriorate through sheer age, whether they are in use or not. A common cause of a blown fuse on an early wireless is a faulty vibrator-converter, this device was used to convert the direct current from the battery to the alternating current required by the set.

Professional overhaul work

There are a few points which are worth noting about the situation which arises when you have work done on your radio by a professional repairer. Firstly, you can hardly expect the repairer to guarantee the entire set when he has only repaired or replaced one or two parts of it. Partial repairs (say £10-£30 inc VAT) can be carried out of course, but any guarantee would be limited to the work done. I have already pointed out that pre-transistor era radios are entirely repairable and that the resulting performance may well be superior to that obtainable from a modern set. Clearly a radio is hardly an assembly which lends itself to partial restoration if the maximum reliability is wanted and although a complete restoration may seem expensive (£100-£150 inc VAT) it will probably prove cheaper and more worthwhile (in the long term) than returning to the repairer at intervals to have a succession of different components replaced or repaired. ☐

The writer wishes to thank Mr Rees of the Vintage Wireless Company (64 Broad Street, Staple Hill, Bristol) for his assistance in the preparation of this article.

Early sets came in all shapes and sizes. This is a 1937 Philco Transitone . . . .

. . . . and this is an American product which employed cable operated remote controls.

"Right then — another tie-breaker . . . Er . . . what year did Harry Belafonte have a hit with The Banana Boat Song?"

Soon after radio sets started appearing in the more well-to-do homes up and down the country in the late twenties, the proud owners naturally started wondering if they could have a set to operate inside their other status symbol, the family car. Fine in theory, but this idea presented a number of problems.

First, there was the question of power supply. Valves require between 100 and 300 volts to operate so how could this be derived from the car's six or twelve volt supply? Then there was the problem of how to make the set compact enough to fit into the car and in a place where it could be controlled from the driving position. Remember how big domestic radios were then compared to now! Finally, a solution to the problem of reception interference from the ignition had to be found. Gradually, however, solutions to the problems were devised and from about 1932 onwards sets were available, though at prices only the wealthy could afford.

To overcome the accommodation problem, many early sets had amplifiers, tuners and power supplies contained in separate 'boxes', while some manufacturers, particularly Americans, produced sets as one large 'box' but with a separate control box connected to the set by Bowden cables. Thus, the main part of the radio could be mounted wherever convenient in the car and the cables run to the control box somewhere on the dashboard. These early sets were very expensive new, but well made and, unlike most modern 'printed circuit' devices, they can be overhauled. Surprisingly enough, the high-street radio retailers seemed slow to realise the potential sales of car radios. In 1939, only 25% of total sales were through radio specialists; all the rest were through the motor trade.

During the war car radios were banned for security reasons. The hostilities, however, provided an ideal opportunity for manufac-

A valve radio that sounds good makes a lovely finishing touch to a period restoration. Adrian James tells how the car radio developed from the domestic type and gives some tips about maintaining them.

Unrestored 1937 Philips radio. The control box is separate from the rest of the radio and connected to it by Bowden cables, enabling the large bulk of the set to be installed away from the cramped dashboard area.

turers to gain experience in building mobile radios for military purposes and, when the first sets after the war appeared, the development was obvious. Valves were now much smaller and many of the sets themselves were more compact though still large by present-day standards, but prices were even higher than they had been in 1939. A push-button Radiomobile cost £27 6s 0d, plus £10 4s 6d purchase tax in 1946 which, incidentally, was the year Radiomobile were formed as a joint venture by HMV and Smiths Motor Accessories. For a considerable period after the war Radiomobile operated a Star dealer network and, for those traders not fully *au fait* with radio installation and the often complex suppression problems, they had a fleet of green vans with experienced fitters who called at garages and carried out the fitting for them.

Exploded view of the same set prior to restoration.

Repairing Radios

Though in-depth repair of car radios requires a fair bit of specialist knowledge and often some equipment there is a certain amount that can be done by the owner using only hand tools. First though, a warning. Though car radios operate from a twelve (or six) volt

One of the first Radiomobiles. The model 100 was made from 1946 to the early fifties, and this one has been completely restored.

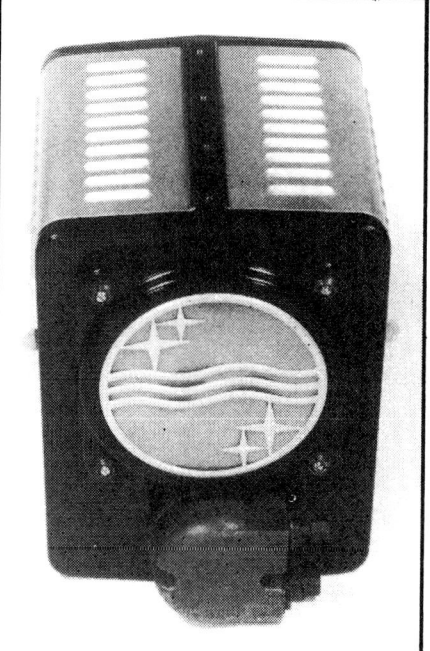

Another Bowden cable controlled set, a 1938 Philips 247B.

obviously to get the thing working, then bring its performance up to scratch by replacing 'soft' valves, condensers that are breaking down etc. Tracking all these faults down and circuit tracing is outside the scope of DIY as it involves special equipment. There is, however, a certain amount of simple fault tracing and rectifying that the ordinary person can do.

Fault
1. *Radio completely dead. ie. no dial light, no sound when turned on.*

Rectification
a. Check that fuse is OK; being the weakest link in a chain they sometimes 'blow' without any apparent reason, but check also that there isn't another fault that has caused the fuse to go.

b. If the set is installed, check that the earth connection is good. Also, if the set has been installed by an amateur, check that there is a proper earth; a set will sometimes work earthing through the aerial, but reception will be affected, and aerial movement may make the earth connection intermittant, leading to intermittant operation. Check this if your set appears not to work when the car is at high speeds, or driving through cross-winds!

c. Check for loose leads, loose connections, plugs in back of set loose, etc., etc.

Fault
2. *Radio dead, no sound, but dial light comes on.*

Rectification
a. Check speaker is OK by substitution, also ensure speaker leads are tight.

b. If radio has a separate amplifier unit, make sure that connections from this to the receiver are sound.

A 1954 Ekco PAP 16, especially designed to fit in a Ford Popular glove compartment.

From the late fifties onwards, transistors started to appear in car radios, initially with valves, but the much smaller size and current consumption from transistors eventually won the day; valves had disappeared completely by about 1964/5.

supply, up to 300 volts is needed across the valves and a vibrator or transformer is provided to step up the voltage. Thus a connected up and switched on valve radio can give a nasty shock to the unwary 'prodder'. So be careful! Many of the radios that I am given to restore are in pretty awful condition. The most important thing is that all the detachable knobs and switches are present; a set's internals can be repaired but, if a knob is missing, the chances are it will be almost impossible to find an exact replacement although in some cases replacements can be made.

Fault finding can be a time-consuming, painstaking business. I find that the worst faults to track down are man-made, ie. where someone else has already had a go and attempted to make the set work. First stage is

WIRED FOR SOUND

c. Disconnect set from battery, remove cover, check valves are properly seated. Remove valves in turn, and clean pins with soft (old!) toothbrush.

Fault
3. *Radio fuse keeps blowing.*

Rectification
a. Check that the fuse is the correct one. Do not be tempted to fit a higher rated one – if the correct fuse keeps blowing there is a fault with the set or its installation.

Fault
4. *Polarity.*

Rectification
Many valve radios were not polarity con-

Badge-engineering in the car-radio world too! This hybrid (valve and transistor) set, badged as an 'His Masters Voice' is actually identical to the Radiomobile 500T, and was made in the early sixties. Plenty of these are still in use.

Radiomobile made especially for Rover P4 series which fitted in the central glove compartment.

scious, part-transistor sets were however. A set that has been connected the 'wrong way' will almost certainly need one or more replacement diodes or transistors, and unfortunately there is no way that this can be detected until the set fails to function and the above tests have been tried.

Fault
5. *Noisy switches and controls.*

Rectification
Crackles and other unwanted sounds when adjusting volume and tone controls are almost always caused by dust or dirt build-up on the contacts. They can usually be elimi-

nated by squirting a drop of WD40 into the 'pot' body. The same trick also improves wavechange switches. Try it if there is excessive crackling when changing waveband, or if one band is significantly quieter than another. Disconnect the set first though!

Parts Availability

Most components for valve car radios can still be found without too many problems; I can supply valves, vibrators etc. in cases of difficulty. The problem, as already stated, is with the trim parts such as knobs and switch buttons etc. When looking for period radios in autojumbles and sales try to find one that is complete. Don't be too concerned if the radio looks tatty; if all the original parts are present this can be rectified.

RELAYS
THE WHY, WHERE AND WHEN

By Joss Joselyn

If you think that all electricity is black magic and the variety used in motor cars is more devilish than most, here is your chance to let in a little daylight. Electrics don't have to be complicated and here we're going to take a practical look at relays and explain both them and their use in simple terms.

A relay is a switch, one that is electrically operated. If you have a relay in the circuit and you close the normal dash-mounted switch, current then flows through the circuit and closes the relay switch — in other words, operates the relay.

Why have a switch to operate another switch? Basically you use a relay when you've added extra components into a circuit or when the switch is used to operate accessories requiring a heavy current. If no relay is used, a larger switch — one with heavy contacts — would be needed to avoid the contacts burning out rapidly. It would also mean a long run of heavy cable if voltage drop, impairing efficient operation of the accessory concerned, is to be avoided. The alternative to all this is to use a normal dashboard switch, normal wiring as far as the relay and then heavier wiring between relay and accessory. If the relay is mounted close to the accessory concerned, it obviously saves a lot of bother.

When do you use a relay? Whenever accessories requiring a heavy current are fitted. Typical examples are a pair of windtone or air horns, a pair of powerful driving lights, heated rear screen or a burglar alarm.

How do you fit a relay? The actual physical work involved is minimal; the majority of what's involved is a matter of wiring. The relay you buy will come together with an integral mounting bracket and fixing is usually accomplished with a pair of large-headed self-tapping screws.

Wiring is a bit more complicated. Some circuits to illustrate what's required for the most commonly used accessories are shown in Figs. 1 to 7. Two relays are employed — the Lucas 33213 which is a four-terminal unit and the Lucas 33188 which is a three-terminal job. In some cases they are interchangeable but this will become clear on studying all the circuits.

It is, of course, possible to use relays of other makes and you can identify which terminals are which from the following table of equivalents.

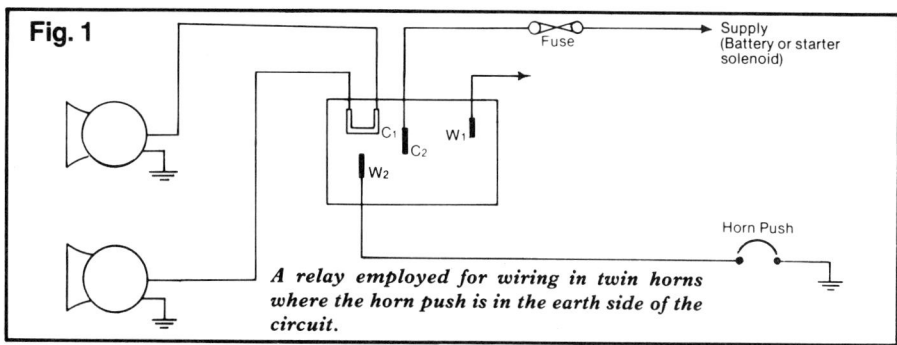

A relay employed for wiring in twin horns where the horn push is in the earth side of the circuit.

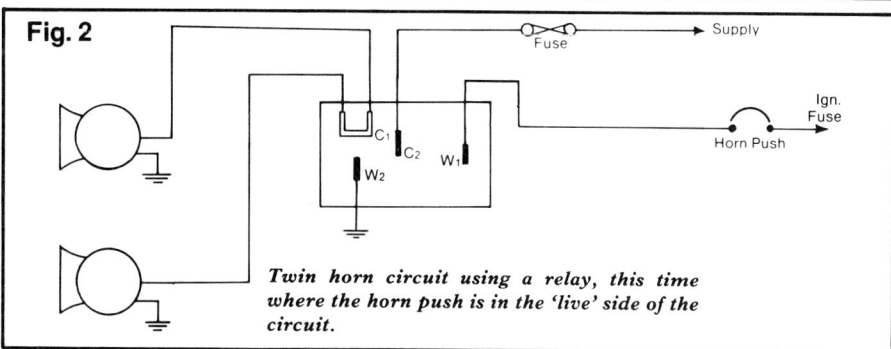

Twin horn circuit using a relay, this time where the horn push is in the 'live' side of the circuit.

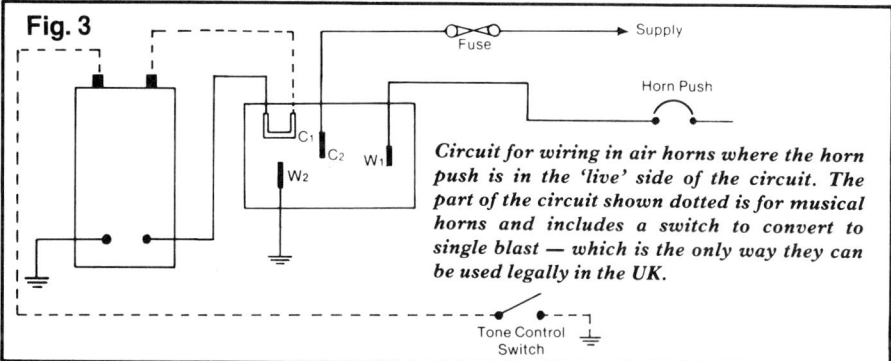

Circuit for wiring in air horns where the horn push is in the 'live' side of the circuit. The part of the circuit shown dotted is for musical horns and includes a switch to convert to single blast — which is the only way they can be used legally in the UK.

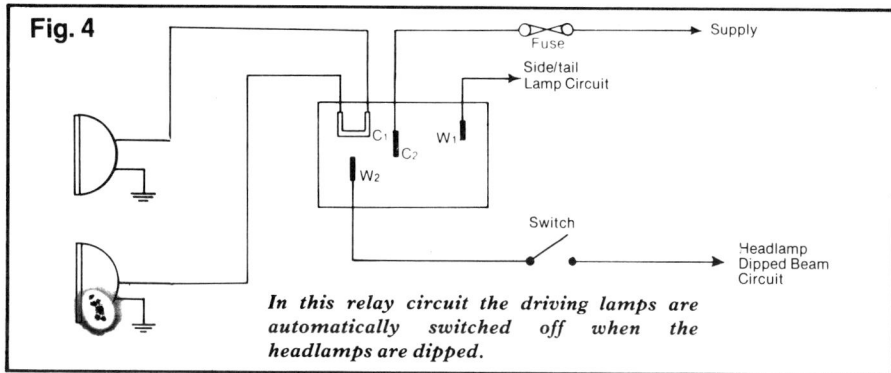

In this relay circuit the driving lamps are automatically switched off when the headlamps are dipped.

Lucas	Bosch	Marelli	Others	
C₁	30/87	B	B	B
C₂	87	H	H	H
W₁	85	PUL	P	S
W₂	86	Unmarked or short lead		

Figs. 1 and 2 show different ways of using relays in the horn circuit. The reason why you might want to do this is if you are changing the original single high-frequency horn for a pair of more powerful windtones or air horns. Whether you use the first or second circuit

(Continued)

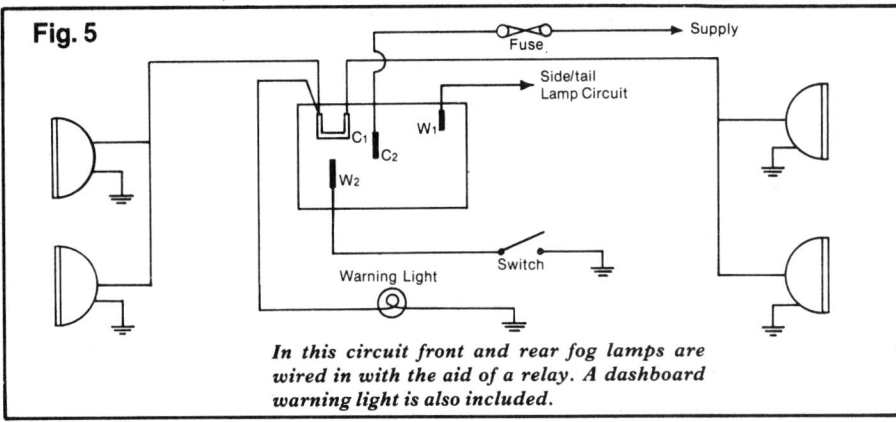

Fig. 5

In this circuit front and rear fog lamps are wired in with the aid of a relay. A dashboard warning light is also included.

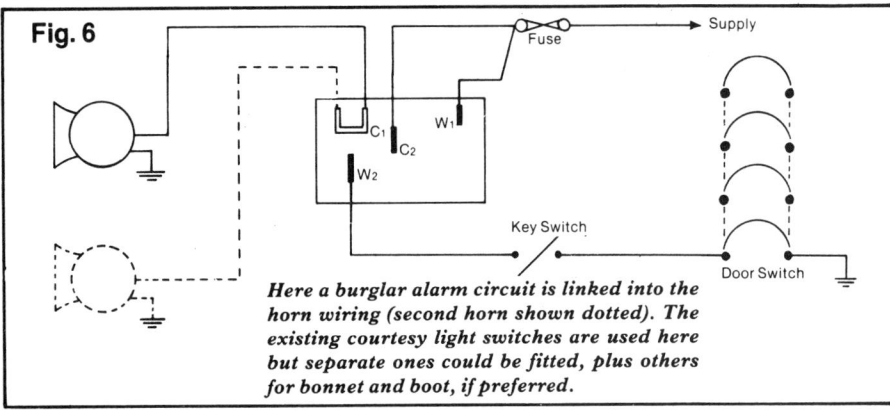

Fig. 6

Here a burglar alarm circuit is linked into the horn wiring (second horn shown dotted). The existing courtesy light switches are used here but separate ones could be fitted, plus others for bonnet and boot, if preferred.

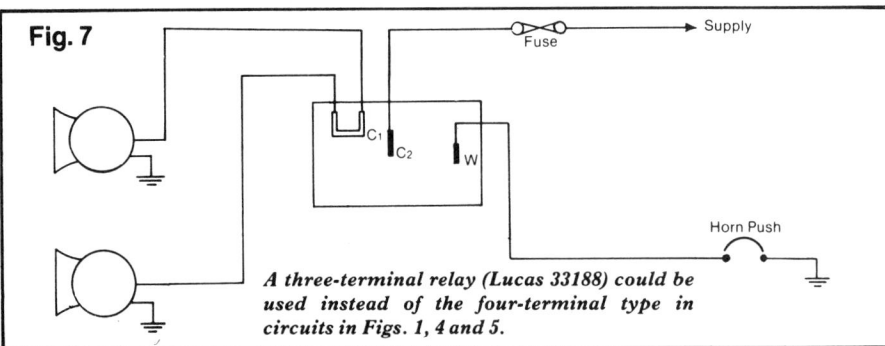

Fig. 7

A three-terminal relay (Lucas 33188) could be used instead of the four-terminal type in circuits in Figs. 1, 4 and 5.

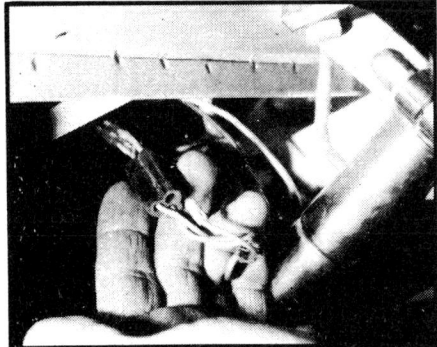

Undoubtedly the simplest way to joint two wires or to join a wire into an existing circuit is to use sleeves and snap connectors.

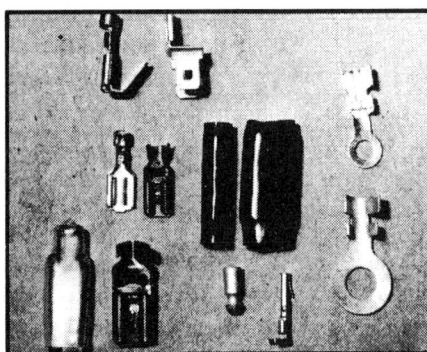

There are many types of terminal that can be used, however. This is a selection of those most commonly favoured.

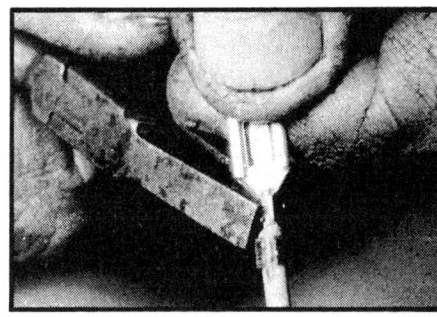

The method of fixing a spade terminal is to thread on the insulated cover first, then to crimp the metal clamp on to the cable insulation, followed by the one on to the wire itself, which is also soldered.

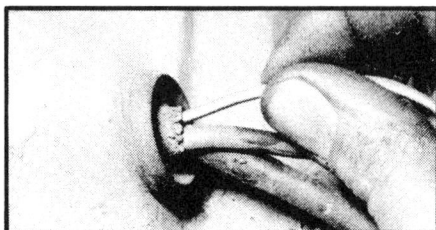

The best way of passing a cable through a metal bulkhead is via an existing hole and rubber grommet. If new holes have to be drilled, ensure they are grommetted to avoid chafing the insulation and causing a short circuit.

Fig. 5 copes with these points.

There are many possibilities with burglar alarm circuits but the one shown in Fig. 6 illustrates the basic principle. In this, when the car is left with the key switch closed, the action of opening a door completes the circuit and sounds the horn(s) which are also connected into the relay. The circuit can be arranged either using the existing door courtesy switches or fitting separate ones.

Fig. 7 illustrates how a three-terminal relay can be substituted for the four-terminal type shown in Figs. 1, 4 and 5.

Wiring Hints

Always use the correct type and size of cable. Buy it from an auto-electrician or car accessory shop and specify 14/30. That is 14 strands of 0.30mm dia. wire. This will cope with the majority of normal work. To carry current to horns or a pair of spotlights, use a heavier cable — 28/.30 (28 strands of 0.30mm wire).

Do not leave lengths of cable draped unsupported. If they are run along the lines of existing wiring, this will be a lot easier and the existing harness supports can be used. If the wiring has to pass through a bulkhead, poke it through existing rubber grommets if possible. If a new hole has to be drilled, always fit a new rubber grommet, otherwise the cable insulation will chafe through and cause a short.

Use proper terminal connections — spade type for joining to the accessories and bullet terminal and snap connector for joining into an existing circuit or for joining wires together. Ensure they are soldered and not just clamped in place and use a good hefty soldering iron — a 100-watt electrical type or a gas iron.

will depend on the car manufacturer's original circuitry — whether the horn push is in the earth side or the live side of the circuit. You may be able to identify this from the wiring in the car or from the circuit in the handbook.

When wiring up driving lamps, you have a couple of legal requirements to satisfy. First, they must not be capable of being lit without the side and tail lamps and, second, they must either go out or be dipped at the same time as

the main headlamps. Circuit No. 4 is the one to use and here in fact they go out when the headlamps are dipped.

Rear fog lamps are now compulsory on new cars and if they are fitted to older ones, they are subject to the same regulations. Without going into too much detail, there too must only be lit in conjunction with side and tail lights and they must also have a tell-tale warning light in the cab. The circuit shown in

ELECTRICAL SOLDERING

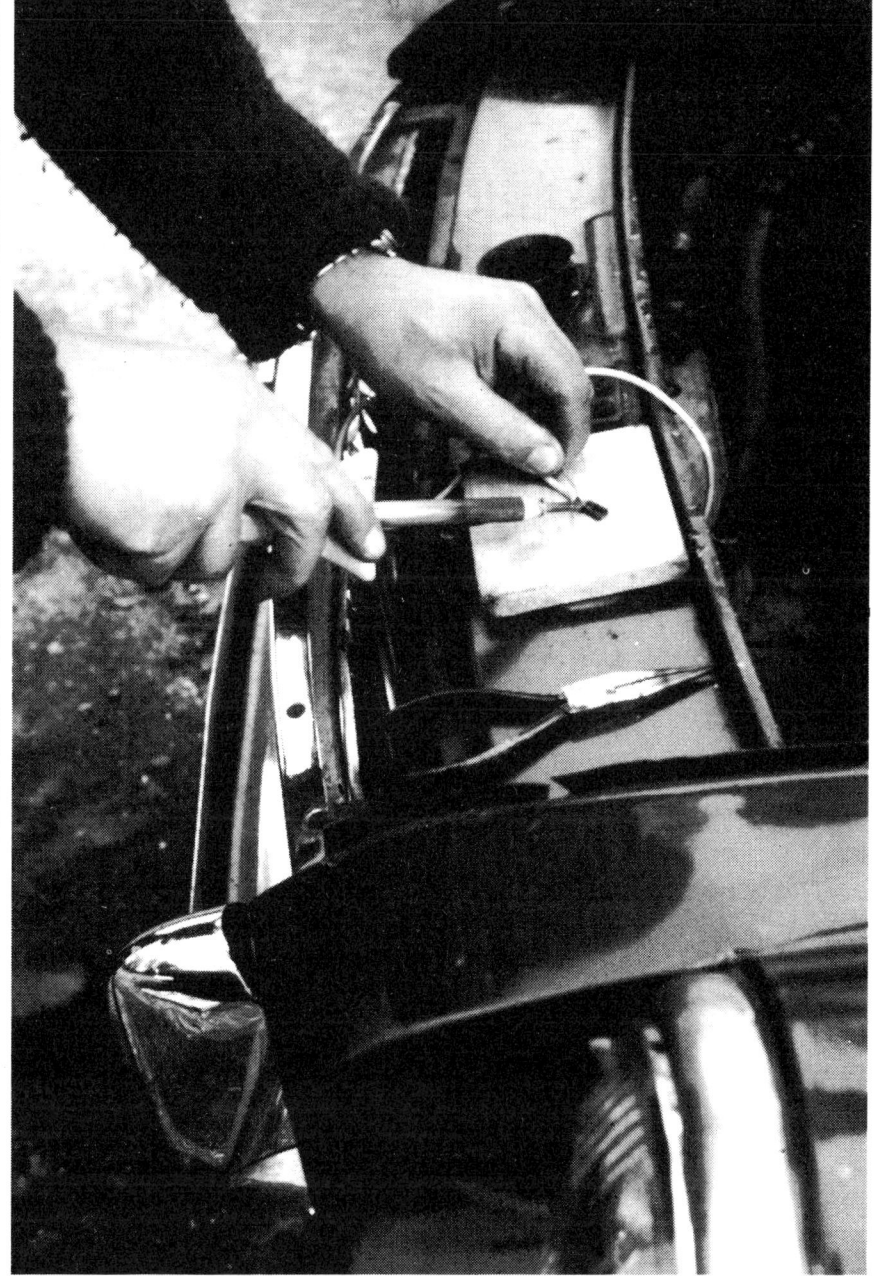

John Williams explains what is involved in this useful process.

Soldering is a method by which two or more pieces of metal may be joined using a molten alloy. The alloy (solder) and the metal objects to be joined are all heated by means of a soldering iron. A flux must be used too, in order to prepare the metal to accept the solder and different fluxes are available for different metals.

Hard soldering employs alloys of copper, silver and zinc and is very similar to brazing. It requires a great deal more heat than soft soldering (using a blowtorch) and is used for making or repairing things which will not require the degree of strength associated with welding.

Soft solder is an alloy of tin and lead and whilst it has a wide range of applications in 'lightweight' repair work its most frequent application in motor cars is in electrical connections, (this article is not concerned with 'body solder', lead loading etc), which illustrate the principles which apply to all forms of soldering.

TOOLS AND MATERIALS

There are electric and flame heated soldering irons. Both types are available in various sizes and if you are buying an iron you should ensure that it will cope with the biggest jobs that you are likely to tackle. Electric irons are much more convenient to use (assuming that a power supply is handy), although when a flame heated iron is in use the blowtorch itself can be used to pre-heat the metals which are to be soldered in certain light construction and repair jobs. A 150 watt electric iron will cope with all but the heaviest cables in a car, but a lot depends upon the conditions in which the work is to be done.

The resin cored solder which looks like silver wire and is sold in a handy dispenser is the type used for electrical connections. Solid bar solder is also available as are other types of flux depending on the metals to be soldered.

HOW IT'S DONE

Soldering is a very simple operation but there are some basic rules which should not be ignored. The metals to be joined must be clean and free from oil and grease and corrosion. When joining freshly bared wires to new terminals this may not be a problem, but in other applications it will usually be necessary to use emery cloth and/or a wire brush to ensure that the areas of metal to which the solder is to be applied will be thoroughly clean. When using bar solder the flux (usually a liquid) is brushed onto the heated metal before the solder is applied. There must be enough heat when the solder itself is applied so that the solder melts to the extent that it will run like water. The metals to be joined should be tinned, that is, given a light coating of solder, and then held in contact with each

The tools and materials required for soldering are few. An electric soldering iron of a suitable size (or a flame heated iron used in conjunction with a blowtorch), some resin cored solder (for electrical connections), emery cloth, pointed pliers, and some insulating tape (if suitable plastic or rubber sleeves are not available).

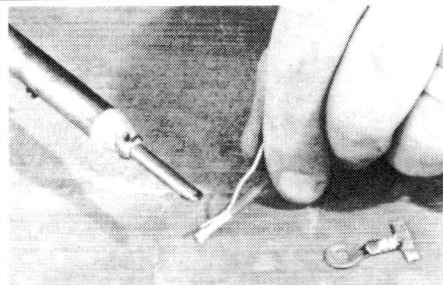

The bared end of the wire is about to be tinned by holding it in contact with a piece of self-fluxing solder with the hot iron. The terminal in the picture has already been tinned.

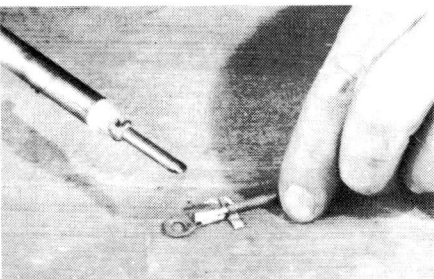

The wire must be fitted to the terminal so that the terminals tabs will fold over and grip the insulation.

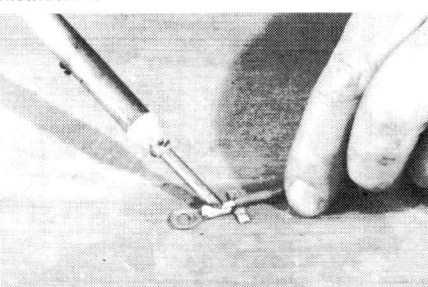

The wire is then held in contact with the terminal using the hot soldering iron — a little more solder can be added at this stage but it should not be needed.

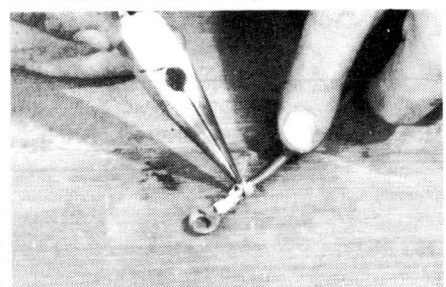

When the solder has cooled the joint can be tested and the terminal clips can be folded over to grip the insulation . . . .

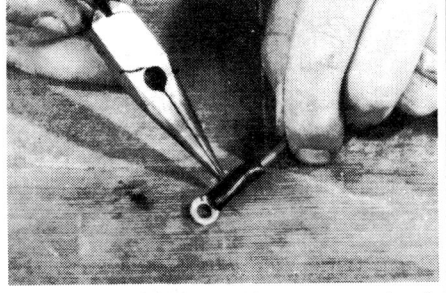

. . . . and a rubber or plastic sleeve (previously pushed along the wire out of the way) can be positioned over the joint.

other (using Mole grips or other suitable clamps) whilst more solder is fed into the joint. Heat is applied throughout this process.

Ideally, soldering should be carried out in a warm still atmosphere to minimise heat losses,

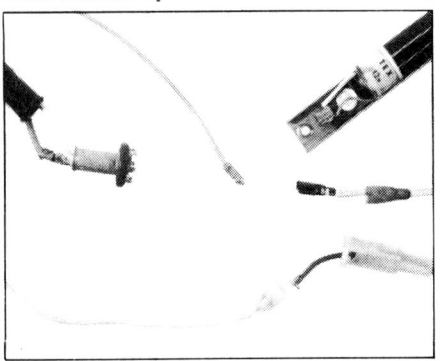

Soldering is the only way to make many of the electrical connections used in cars, and where solder is not strictly necessary it can often be applied to make a better electrical connection and a longer lasting join.

and the work surface used should be of a material which will not conduct heat away from the components being soldered. For small soldering jobs a piece of scrap wood (as used in our pictures) makes a good bench, and when soldering larger pieces of metal it may be desirable to wrap the components to some extent in a heat insulating material to reduce heat losses. In all cases precautions should be taken to avoid any risk of fire. With a little practice soldering can be carried out quickly and efficiently and this will be necessary when working on some electrical components. Prolonged application of heat can cause damage elsewhere due to heat conducted away from the soldered area. Sometimes Mole grips or similar can be attached to the component to intercept some of the conducted heat between the area being soldered and the part which may be damaged by heat, but often it is a matter of working quickly; the solder should also be prevented from running where it is not wanted.

ELECTRICAL FAULT FINDING

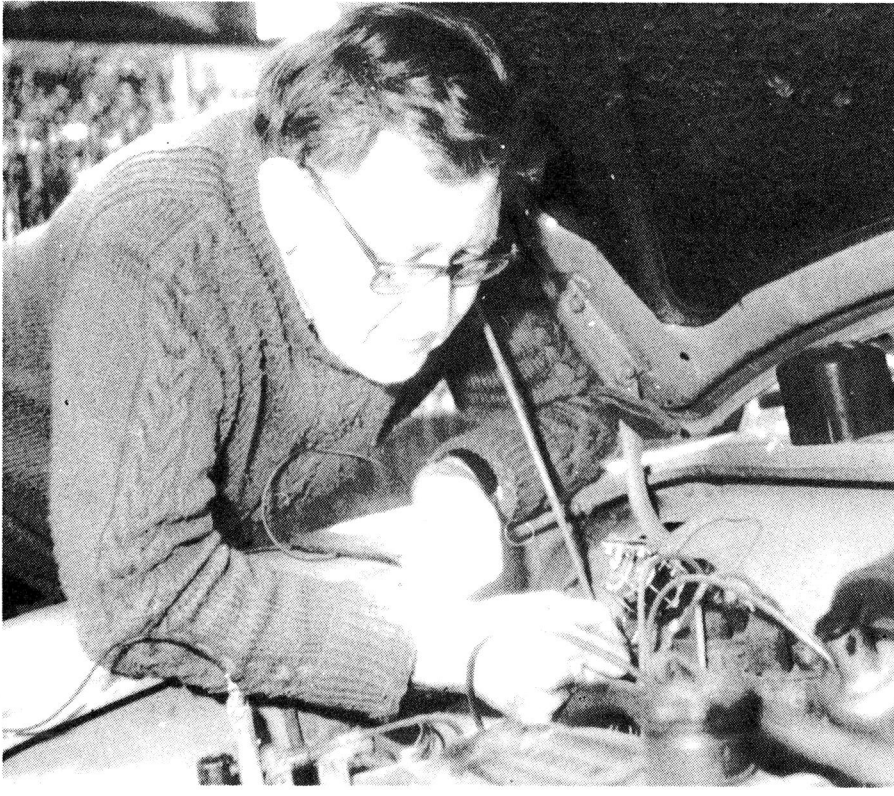

Not sure you know how? Joss Joselyn answers some of the questions you never liked to ask.

'**O**n the blink' is a phrase that has come to mean 'something is going wrong' and it is no accident that it seems to have electrical overtones. The motor car is a complicated mix of both mechanical and electrical components and, while mechanical faults can be extremely frustrating and difficult to trace, it is the elusive electrical defect that sends most of us right round the twist!

The trouble with electrics is that they *look* complicated — a lot more complicated than they actually are. The cables resemble multi-coloured spaghetti and anyone who thinks he'll consult the circuit diagram as a solution ends by wishing he'd taken up nuclear physics or brain surgery instead!

But it can all be explained — honest!

That coloured cable insulation does actually have a pattern and a meaning. Not all cars use the same system, but fortunately many do use the British Lucas system which has applied to nearly all British cars from around 1950.

It is based on seven colours — black, brown, white, red, blue, green and purple. Each base colour indicates a different section of the wiring. Cables in the charging circuit together with feed wires from the battery to main switches are brown. White denotes the ignition circuit. Blue is headlamp wiring. Red is side and rear light wiring. Auxiliary or accessory circuits, controlled via the ignition switch, have green as their base colour and those not ignition controlled, purple. Black is the colour for earth wires. There are one or two more colours which appear on later cars that are particularly well endowed with accessories — light green, yellow, slate and pink.

Each base colour, as it works its way through the individual circuit, is modified by means of a second 'trace' colour. This is normally a thin stripe along the length of the cable, and the colour changes depending on where in the circuit the cable is. The trace changes, for instance, following a fuse or a switch.

Although the theoretical circuit contained in the car handbook or workshop manual is likely to bear little resemblence to the actual wiring in the car (it's probably drawn theoretically, rather than practically), the colour coding will be the same. Letters on the circuit indicate the wiring colours as follows:

B Black
G Green
K Pink
LG Light green
N Brown
P Purple
R Red
S Slate
U Blue
W White
Y Yellow

The main diagram shows a typical basic circuit. It should be possible to trace all the circuits, identifying the colour changes in the wiring as you go, but a few comments might make it easier.

Start at the battery. The normal two thick cables are there — the earth and the one to the solenoid. From here, one thick cable powers the starter motor but there are also a couple of brown cables, one carrying power to the control box and the other to the fuse box. The wire to the control box goes via an ammeter and it will be noticed that all the wires to this component have the charging circuit brown base.

Two of the wires leaving the control box are brown/blue and they feed the ignition switch and the lighting switch. From the ignition switch the wires that lead off are ignition white. Those from the lighting switch assume the base colour blue, blue/white to the main beam circuit and blue/red the dipped beam. In addition a plain red cable feeds the side and rear lights and the panel light switch. A red/white cable then feeds the panel light bulbs.

Plain green wires lead out from the fuse box to feed the ignition controlled accessories. One goes to the stop lamp switch and from there to the actual lamps via green/purple wires.

The flasher unit is fed by another plain green wire and it is linked to the indicator switch by a light green/brown cable. A green/white cable feeds the offside indicators and green/red the nearside. The other ignition controlled accessories are all fed in the same way.

There are fewer accessories not protected by the ignition switch — usually only the horns and courtesy lights. A purple wire with no

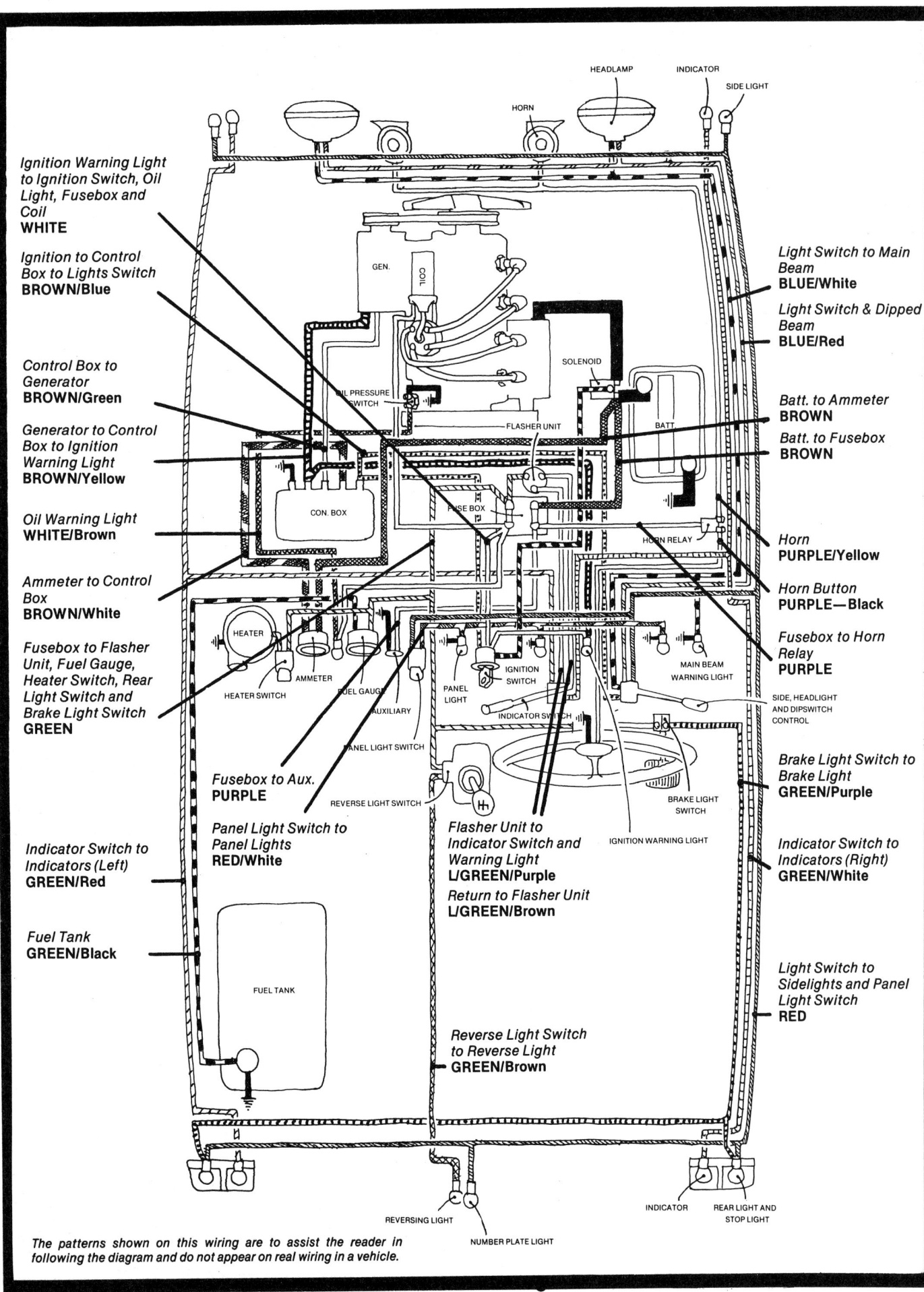

Ignition Warning Light to Ignition Switch, Oil Light, Fusebox and Coil
WHITE

Ignition to Control Box to Lights Switch
BROWN/Blue

Control Box to Generator
BROWN/Green

Generator to Control Box to Ignition Warning Light
BROWN/Yellow

Oil Warning Light
WHITE/Brown

Ammeter to Control Box
BROWN/White

Fusebox to Flasher Unit, Fuel Gauge, Heater Switch, Rear Light Switch and Brake Light Switch
GREEN

Indicator Switch to Indicators (Left)
GREEN/Red

Fuel Tank
GREEN/Black

Fusebox to Aux.
PURPLE

Panel Light Switch to Panel Lights
RED/White

Flasher Unit to Indicator Switch and Warning Light
L/GREEN/Purple

Return to Flasher Unit
L/GREEN/Brown

Reverse Light Switch to Reverse Light
GREEN/Brown

Light Switch to Main Beam
BLUE/White

Light Switch & Dipped Beam
BLUE/Red

Batt. to Ammeter
BROWN

Batt. to Fusebox
BROWN

Horn
PURPLE/Yellow

Horn Button
PURPLE—Black

Fusebox to Horn Relay
PURPLE

Brake Light Switch to Brake Light
GREEN/Purple

Indicator Switch to Indicators (Right)
GREEN/White

Light Switch to Sidelights and Panel Light Switch
RED

HEADLAMP   INDICATOR   SIDE LIGHT

HORN

GEN.   COIL

OIL PRESSURE SWITCH

FLASHER UNIT

SOLENOID

BATT

CON. BOX

FUSE BOX

HORN RELAY

HEATER

AMMETER

HEATER SWITCH

FUEL GAUGE

AUXILIARY

PANEL LIGHT SWITCH

PANEL LIGHT

IGNITION SWITCH

INDICATOR SWITCH

MAIN BEAM WARNING LIGHT

SIDE, HEADLIGHT AND DIPSWITCH CONTROL

BRAKE LIGHT SWITCH

REVERSE LIGHT SWITCH

IGNITION WARNING LIGHT

FUEL TANK

REVERSING LIGHT

NUMBER PLATE LIGHT

INDICATOR   REAR LIGHT AND STOP LIGHT

The patterns shown on this wiring are to assist the reader in following the diagram and do not appear on real wiring in a vehicle.

54

ELECTRICAL FAULT FINDING

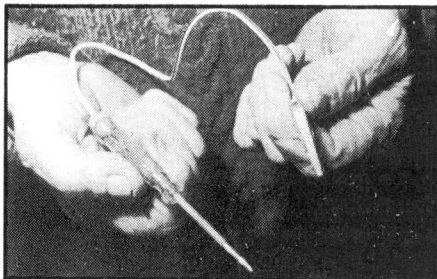

Typical professional circuit tester, available quite cheaply from any motoring accessory shop.

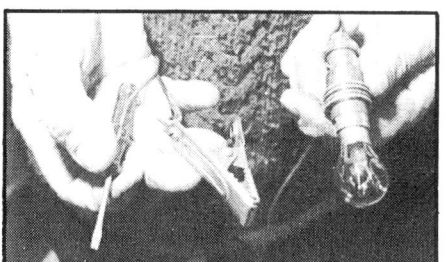

A home-made equivalent. The two 'terminals' are on the left and instead of using the screwdriver a permanent probe can be soldered to the end of the wire.

(Continued)

trace feeds the switches of both horns and courtesy lights. Between switch and horn, it is purple/black and between switch and courtesy lamp, purple/white.

Once the colour coding has been licked into some sort of sense, the whole thing can be made a lot simpler by separating every component in the car into a single simple circuit. The basic circuit will include the battery, a switch and a consumer unit, say a light or a motor, and the wiring to connect them all together. There may also be a fuse but this is best dealt with separately.

For the purpose of illustrating a logical and basic test sequence, let's assume, for instance, a light has stopped working. To find the problem you will need a simple circuit tester, either a professional one available from any accessory store or a home made one which can be a bulb holder with two wires attached. This latter can either be cut from an old car wiring harness or made up from scratch, but generally it is easier to buy a proper tester from the shop. A 'jump lead' — a length of wire with a clip either end — will also be found useful.

Test the tester itself first by clipping one terminal to one side of the car battery and touching the other to the second terminal. If the battery is working, so should the test lamp bulb. If it doesn't light, check it on another battery.

If the bulb did light on the first check, leave the crocodile clip connected to point 1 and touch the probe to point 3 on the switch. If it lights, there is power to the switch.

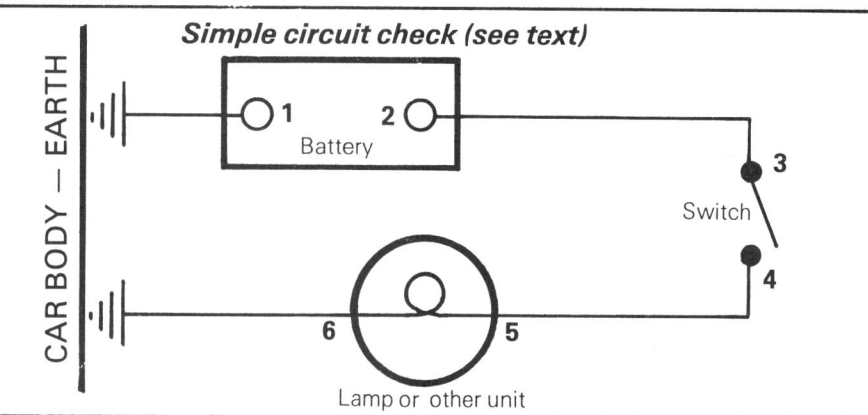

Simple circuit check (see text)

CAR BODY — EARTH

Battery — 1 — 2
Switch — 3 — 4
Lamp or other unit — 6 — 5

Close the switch and check at point 4. If the switch is OK, the bulb will light. If no fault is shown, move on to point 5 on the consumer unit. If this lights the bulb, move on to point 6. Because both point 1 and point 6 are at the same potential, the bulb won't light, so move the crocodile clip across to the other battery terminal (point 2) and this will check the last part of the circuit.

If the bulb lit all the way round the circuit but the consumer unit still does not work, it must be the unit itself which is faulty — blown bulb or burnt out motor or something.

Logically, you know directly the test bulb fails to light on your way round the circuit that the fault lies between that test point and the previous one where the light did show. OK, but what sort of fault?

Between components it is either that a connection has come apart or the wire has become damaged or severed or corrosion has caused a poor connection. Checking means pulling apart snap connectors, cleaning the bullet ends and if the inside of the sleeves is rusty, fitting a new one.

Once pulled apart the 'bullet connectors can be cleaned up with emery cloth. The sleeve part is best replaced if at all suspect.

This sort of rust inside a lamp can very easily cause a poor earth and prevent the bulb from lighting.

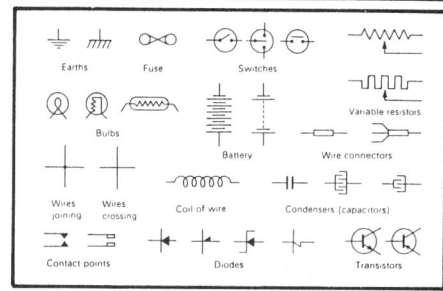

These are some of the symbols used in circuit diagrams to denote particular components.

Cable damage usually means that the wire has strayed near a hot exhaust, become chopped up by a moving component, like the radiator fan, or has been trapped or crushed during repair work — under the radiator or the bonnet, for instance.

If the problem appears to be the switch, i.e. current doesn't pass through it when it is closed, you can check by by-passing it with a plain wire. If the component in the circuit then works, you can use this wire as a get-you-home repair, although it does mean the component will be 'on' permanently.

You will already know that the car chassis is used as the 'earth return' to complete every circuit and thus every circuit is eventually connected to earth. Sometimes separate 'black' wires are used for this but often the earthing is through the mounting bolts and the casing of the component itself. This applies particularly with lights and, because they are often sited down in the path of mud and wet from the road, it is not unusual for dirt and corrosion to interfere with the electrical contact and for the unit to stop working.

Restoring the earth is basically a cleaning up job, or, if the lamp casing is badly rusted, perhaps fitting a replacement. Check and clean the contact between the side of the bulb and its holder, between the bulb holder and the lamp body and between this and the metal of the car.

Although the theoretical principles outlined are valid for every circuit in the car, it must be realised of course that they may well be a bit more complicated and a bit different. Each is not connected directly to the battery and there may well be more than one consumer unit in each circuit — like the side and rear light circuit, for instance, which has four lamps.

It is also a good idea at this stage to mention fuses. One fuse may well cover several circuits and switching on other components protected by the one fuse is often a quick way of checking whether there is a fault or if it is a blown fuse that is the problem.

There are plenty of other short-cuts too which can save a lot of checking work. Staying with the sidelights circuit, if, for instance, the side lights and the panel lights work, but not the rear lights, the fault cannot be in the feed to the switch or the switch itself.

Try the fuse first, as some cars have the rear lights separately fused. If this brings them back on, it might just have been a faulty fuse, but it is worth while checking through the rear lamps circuit for loose connections etc. anyway.

If the fuse blows again instantly, try disconnecting the rear lamps and then fit another fuse. If all is well, reconnect the lamps again one at a time to find which has the fault and then locate it. If the fuse still blows, the fault is between the fusebox and the point where you disconnected the rear lights.

If that lighting circuit is not fused, and a lot of them aren't, the problem is a break in the circuit somewhere or a faulty connection. Then you can use the routine already described to check through the faulty rear part of the circuit.

Similar commonsense checks are possible on most circuits, once you know how they are wired and fused. There are also additional quick checks which are possible on other circuits. Let's look at one or two of these.

Direction Indicators

If the indicators on one side work normally and there is a problem on the other, logically, the fuse is OK and so is the power supply to the flasher unit. The problem could be the switch (if neither light works) and this can be checked with the test lamp.

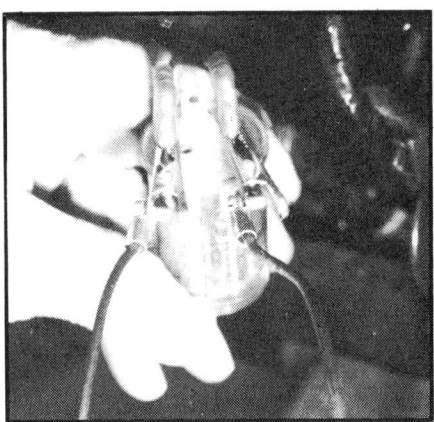

Without disconnecting the flasher, the two connections are joined by means of a small 'jump' lead. If the lights come on but don't flash, the flasher unit is faulty and should be changed.

If the lights on one side work quicker or slower than usual, the probability is that a bulb has gone or there may be a partial short circuit or perhaps a high resistance connection to one of the bulbs.

If all the indicator lamps stop working, check the fuse first. If this does not help, pull the green and green/brown wires off the flasher unit and join them together. Operate the switch and the indicator lights on that side should come on but not flash. This will point to the flasher being faulty. If there are still no lights, the probability is that the input wire to the switch has become disconnected or broken off.

Charging Check

First signs of charging trouble may be if the charging light comes on and stays on. Most likely this is a broken fanbelt but if not, it means a faulty dynamo. Run the engine at a fast idle and manually close the cut-out contacts in the control box. A lot of bright blue flashes from the dynamo will indicate worn brushes and commutator.

If the cut-out opens again when you release it, it's probably more serious dynamo trouble.

Now try bridging the regulator contacts with a screwdriver and if this causes the cut-

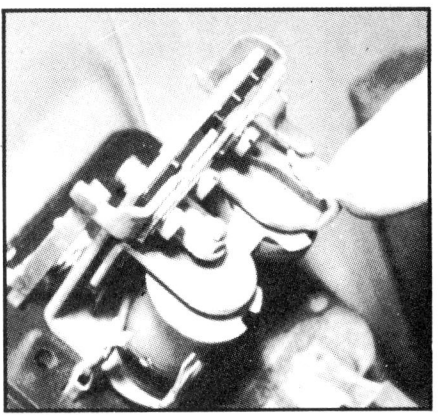

Manually closing the cut-out contacts in the control box.

Bridging the regulator contacts with a screwdriver.

out contacts to close, the trouble is in the regulator contacts.

If there is no warning light at all, the bulb may have blown but if not, you'll probably have to fit a new control box.

If the warning light gets brighter the faster you run the engine, there may be problems in the cut-out or the control box earth connection has broken or is corroded. This latter can be repaired.

Probably the simplest thing you can do to keep an eye on the charging system is to fit either an ammeter or a voltmeter (battery condition indicator).

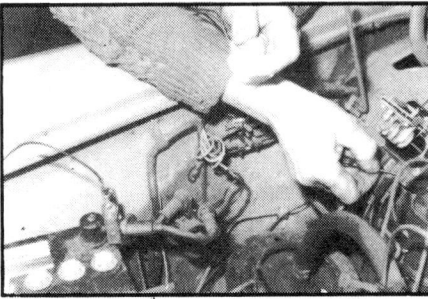

Using the simple bulb tester to check earth continuity of the regulator unit. One clip goes on the battery live terminal, the other to the earth terminal on the regulator. The test bulb should light.

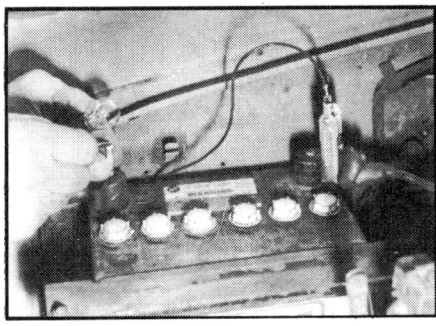

Possibly the simplest check of all, one end of the fuse goes to one battery terminal, the test bulb contact touches to the other end and the second test bulb contact clips to the other battery terminal. If the fuse is sound, the bulb will light.

Two final simple tests, one using a circuit tester of the type with a battery incorporated. Using this is the simplest way to check continuity through a doubtful fuse — they aren't always easy to see.

The other check is for an earth fault. If this is suspected for instance in a lamp unit which isn't working, switch the lamp on, attach one end of the jump lead to a good clean earth point and dig the other clip into the metal of the lamp, into the bulb holder and the side of the bulb. If any of these jabs cures the faulty lamp, bad earthing was the trouble.

I've only described here some of the simpler aspects of electrical fault finding but luckily many of the faults that occur are simple things anyway. Wielding a simple circuit tester, it is surprising what you can discover and even more surprising what the tests can mean if you apply a little common sense.

Wiring-

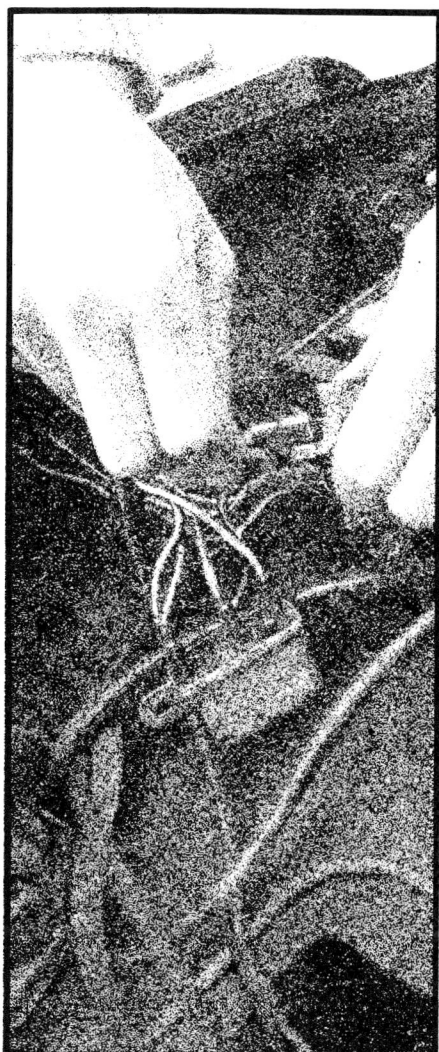

Fitting A New Loom

Could you do-it-yourself? John Williams investigates what is involved.

If there is one part of a car which will often baffle even a professional mechanic it is the electrical system. Although many owners are able to repair local electrical faults very few will tackle a complete rewiring job, so when the Practical Classics Riley 1.5 was re-wired I went along to see what it involved.

A new wiring loom is likely to be needed in the course of a complete restoration if only for the sake of appearances, but it is also needed if the existing loom is showing signs of serious deterioration due to age, or evidence of damage due to fire or excessive heat (often caused by welding too close to a section of the loom). Extensive 'repairs' or modifications carried out by previous owners can also leave the original loom in such a bad state — probably with redundant wires and loose ends throughout the car, and the 'new' wiring bearing no relation to the wiring diagram — that a new loom may well be the easiest, if not the only, answer to the problem.

The New Loom

If the original wiring loom is no longer obtainable a replacement can be made to order for most cars built in the last twenty years, and our new loom was supplied by Riley specialist John Foster (0924 409329). If it is not possible to find a supplier or a one-make specialist who has the necessary pattern to make a new loom, it is worth bearing in mind that any competent auto-electrician should be capable of producing a new loom, using either the old loom or the car itself as his guide.

If there is a choice of suppliers for the loom you require it will be worth finding out exactly how much wiring will be supplied; you may also be able to choose between plastic and fabric covered looms. Some suppliers claim to provide every piece of wiring to be found in the car whilst others supply the main part of the harness (which occupies the engine compartment and the area behind the dashboard) and the section which extends to the boot but do not supply some of the other wiring required. On our Riley 1.5 for example the wiring which may not be supplied includes the short section from the dipswitch to the main loom, the section which passes through the bootlid, and the section which passes around the perimeter of the boot to feed the nearside rear lights. What is, or is not supplied will vary from car to car and will depend to some extent on what was considered part of the wiring harness in the first place and on what information is now available to the harness maker. The extra sections of wiring which may not be supplied are likely to be simple and easily made up by the owner who can certainly save money by doing this rather than pay for a professional's time.

Unless the harness maker is able to inspect the actual car for which he is making a loom he cannot be expected to know which connectors will be required for every component. Where different branches of the loom are joined directly to each other this is achieved by means

The Riley 1.5 front harness as supplied by John Foster with a packet of snap connectors and (coiled) the rear section of the loom which passes along the edge of the roof.

of nipples and snap connectors on many cars but it is individual components such as the control box, solenoid, dynamo, temperature sender unit etc which create problems. Older examples of some of these components had built-in screws under which were secured the bare ends of the wires, whereas their recent counterparts rely upon Lucar ('spade terminal') connectors. Similarly, for older dashboard instruments, ignition coils etc the wiring would have to be fitted with eyelets which were secured to each threaded terminal on the component by a small nut and these too may well have been replaced and now require Lucar terminals. Replacement wiring harnesses made to original patterns tend to include just enough wiring to reach the various components of the electrical system and since changing connectors will inevitably cause wires to be shortened this can create problems. This is done on the premise that longer wires add to the cost of the loom and if the connectors as fitted do not require changing the installed wiring will be untidy. However, I would like to see an extra inch or

Before fitting the new loom you will probably need to buy some grommets to replace those which are damaged or missing and a supply of Lucar (or 'spade') terminals to replace some of the ringlet type terminals fitted to the harness in accordance with original specifications.

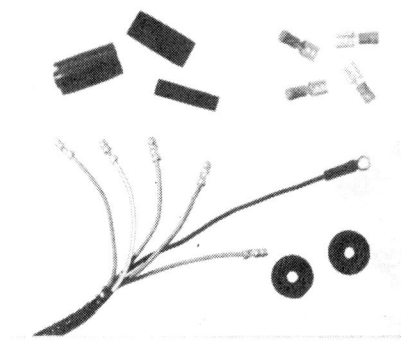

Wiring- Fitting A New Loom

(Continued)

two on each wire in the engine compartment, which would add little to the cost of the loom — wires which are too long can easily be shortened.

Re- wiring Our Riley

The main reason for re-wiring the Riley at this time was to eliminate the many electrical problems which have made the car so unreliable in recent months.

On our Riley 1.5 the wiring can be considered in four sections. Firstly there is the front part of the harness which extends from the dashboard instruments through to the engine compartment. To this is attached by means of nipples and snap connectors the rear harness which passes up through the offside windscreen pillar along the edge of the roof and down into the boot. Then there are those wires which feed such components as the dipswitch and gearbox–mounted reversing lamp switch and connect the rear harness to the nearside rear lights, and finally the wiring which is part of the front and rear light assemblies.

The rear part of the loom (which passes through the roof) was not faulty and as it appeared (at both ends at least) to be in a fair condition we decided to leave it alone.

On some cars it is possible to attach the new 'roof section' of the harness to the old wiring and use one to pull the other into place. On others (including, we suspect, the Riley) the harness is secured to the roof by clips and it is necessary to release the adjacent headlining. We were also lucky in that although there had been a number of problems affecting the headlamps no faults were apparent in the wiring attached to the lamp units. It was clear however, that some of the wiring in the boot and bootlid would need replacing together with the front half of the loom.

Removing The Old Harness

The work on our Riley was carried out by Mr Holloway of West Wickham (telephone 01-777 7210). The fist job was to disconnect the battery. Then the old harness was cut at the

The battery must be disconnected first and it should be cleaned together with the battery posts and new terminals and earth strap fitted when the re-wiring is completed.

Tools you will need

Spanners (for some earth lead bolts) • **screwdrivers** • **pliers** • **inspection lamp** • **releasing fluid** • **jack and axle stands** • **Lucar terminals** • **fuses** • **insulating tape** • **battery terminals and earth lead and extra wiring of a suitable diameter and colour.**

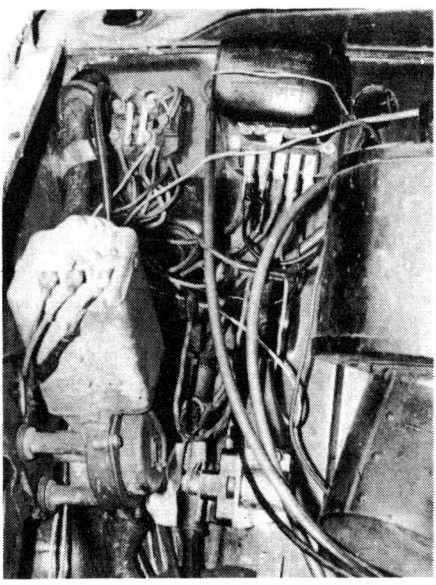

There will not be enough space to install the new harness as the old one is disconnected (left), the colours of the old wiring may be indistinct and difficult to compare with the new harness even if the old wiring is complete and unmodified. Having removed the old wiring (right) take the opportunity to clean those components and areas of the engine compartment which were previously obstructed, fit new fuses in the fusebox and new grommets to replace those which are damaged or missing.

bulkhead so that the wiring in the engine compartment and under the dash could be extracted separately. Wherever nuts or bolts were used to secure wires these were replaced where they belong so that should one of them would be lost, no time would be wasted sorting them out later. On the old wiring which was to remain in place (to the lights and the rear section of the loom etc) the nipples were checked to make sure that they were firmly attached to the wires, and new snap connectors were fitted ready for the new loom.

Fitting The New Wiring Loom

The first job was to feed the dashboard section of the loom through the bulkhead from the engine compartment taking care to avoid damage to the wiring or the instrument light fittings which come fitted to the loom. Then the engine compartment wiring was laid out to check how well it would fit, taking into account any loss of wiring where connectors would have to be changed. It was realised at this stage that the new harness included wiring for twin horns and twin spotlights, but our Riley has a single horn and no spotlights.

This table indicates the most common system of colour coding used in the wiring of British cars. Certain Fords and many foreign cars follow different systems.

Circuit	Colour				
Battery or solenoid switch to ammeter, if fitted	Brown	Dynamo 'D' terminal to control box	Brown/yellow	Ignition fuse to wiper motor	Green
Battery or solenoid switch to control box (dynamo)	Brown	Dynamo 'F' to control box	Brown/green	Ignition fuse to flasher	Green (but wh
Battery or solenoid switch to switches (bu-passing control box)	Brown	Ignition switch to coil SW or + terminal	White	Ignition fuse to stop-light switch	Green on Imp and
Battery or solenoid to alternator	Brown	Ignition switch to ignition-controlled fuse	White	Ignition fuse to heater fan motor	Green Herald)
Ammeter to control box (dynamo)	Brown/white	Ignition switch to petrol pump	White	Ignition fuse to reversing-light switch	Green
Ammeter to switches	Brown/white	Ignition switch to ignition warning light	White	Ignition fuse to instruments	Green
Control box to switches (dynamo)	Brown/blue	Ignition switch to oil light	White	Light switch to side and tail lamps	Red
		Ignition switch to accessory fuse	White/blue	Light switch to side and tail fuse	Red
		Ignition switch to starter solenoid	White/red	Light switch to panel-light switch	Red

Extra wiring (preferably of the correct diameter and colour) will be needed to make up those sections not supplied with the new harness, one of which on the Riley 1.5 was the cable to the dipswitch (arrowed).

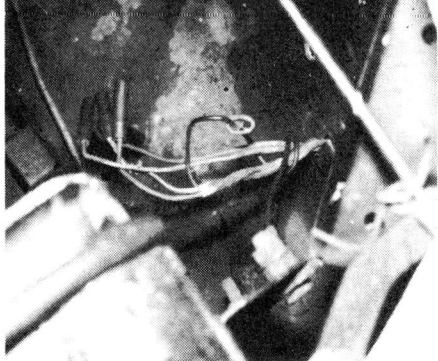

In many cases the wiring attached to lamp units is an integral part of the unit and cannot be replaced separately but check that it is working satisfactorily, that the insulation is sound, and the nipples firmly attached.

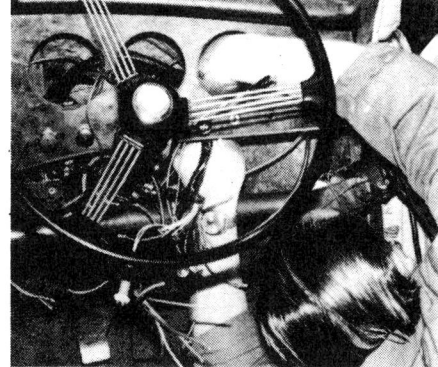

New dashboard wiring should be supplied complete with the instrument light fittings and the old wiring should be complete with as many switches as possible so that these can be transferred directly to the new loom, thus reducing the number of connections which have to be made (right) when the loom is behind the dashboard and access is likely to be limited.

When the harness is laid out in this way the bundles of wires which branch out at intervals will be seen to be near the components which they serve and the aim is to finish with a neat job where the harness is adequately supported but there is no tension in any of the wires.

The dashboard wiring presented the most problems although it must be said that Mr Holloway made the whole job look remarkably, if deceptively, easy. The loom passed through the bulkhead on the offside of the car and was so designed that immediately behind the bulkhead there was a bundle of six wires fitted with nipples for attaching to the roof section of the harness. Then there was a plain section of harness which had to be fitted across the area behind the dash towards the centre of the car and then looped back upon itself so that the branches of wires could be fitted where they belonged, the end of the loom being looped back again to the centre of the car to feed the heater. In other words there was well over two feet of a fairly thick section of the harness (containing about nineteen wires) which had to be accomodated behind the dash taking up valuable space before the instruments and switches could be connected.

In the boot it was necessary to make up new wiring to connect the nearside lights to the roof section of the loom, and we renewed the wiring through the bootlid to the number plate/reversing light assembly, and this assembly will have to be replaced (if we can find a replacement) on account of internal damage. The wire from the fuel tank unit (which passes through the boot) was found to have about an inch of insulation missing — which no doubt accounted for the strange behaviour of the fuel gauge from time to time.

When the re-wiring was completed new terminals were fitted to the battery leads, the battery connected, and it was time to discover whether everything worked normally. The offside front indicator refused to flash but the sidelight obliged instead. Mr Holloway used a spare piece of wire to earth the lamp unit to the front bumper, and the indicator worked. He then connected the lamp to the side grille in which the lamp unit is mounted, and the indicator failed again, proving that there was no satisfactory electrical contact between the offside grille and the body. This was remarkable in view of the number of attachment points for the grille, but on tightening one of the screws which attach the grille to the body the lights worked normally again. There was also an absence of any electrical contact between the lamp assembly and the bootlid to which it is attached by three stout screws. Repeated attempts to clean one of the screws and the bootlid adjacent to it failed to produce a result. Our efforts were transferred to a different screw with

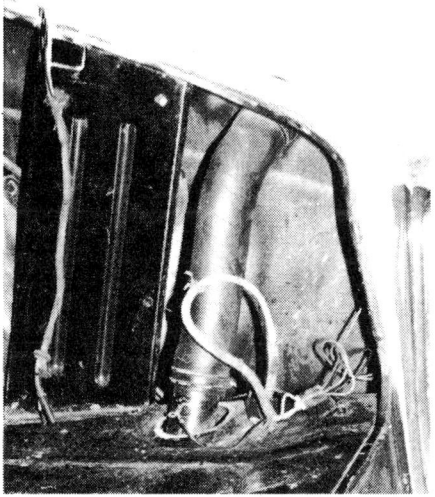

If the main loom was in poor condition other circuits will probably need re-wiring too including the reversing lights and number plate lights and fuel tank sender unit.

Panel-light switch to panel lights	Red/white	Stop-tail switch to stop lights	Green/purple (Ford: green/ yellow)	Oil warning light to transmitter unit	White/brown
Side and tail fuse to N/S side and tail lamps	Red/black or red/brown	Reversing-light switch to reversing lights	Green/brown	Ignition warning light to dynamo 'D' terminal, or alternator 'IND'	Brown/yellow
Side and tail fuse to O/S side and tail lamps	Red/orange or red/brown	Flasher unit to indicator switch	Light-green/ brown	Coil to distributor	White/black
Side and tail fuse to side and tail lamps, both sides	BL: red/green Vauxhall: red/blue	Flasher unit to indicator warning light	Light-green/ purple	Fuse to horn	Purple (brown if not fused)
Lighting switch to dipswitch	Blue	Indicator switch to N/S flashers	Green/red	Fuse to interior light	As above
Dipswitch to dip beam	Blue/red	Indicator switch to O/S flashers	Green/white	Fuse to clock	As above
Light switch to main beam	Blue/white	Petrol gauge to tank unit	Green/black	Horn push to horn	Purple/black, brown/black
Wiper switch to motor (wound-field)	Black/green	Temperature gauge to transmitter unit	Green/blue	Horn relay to horn	Purple/yellow
Wiper switch to motor (permanent-magnet)	Blue/green, red/green, brown/green			Interior light to door switch	Purple/white
				Earth	Black

Wiring- Fitting A New Loom

(Continued)

Earthing faults may be the most troublesome after the re-wiring is completed. Our offside indicator was found to earth to the side grille but not to the bodywork . . .

. . . and the number plate/reversing lamp assembly was not earthing to the bootlid, despite the fact that it was attached by three screws.

immediate success, but the problem was bewildering while it lasted and I began to wonder whether the fault was elsewhere.

Could You Do It Yourself?

If by virtue of your knowledge of the electrical system you could re-wire a car from scratch you should have no difficulty with a ready-made harness which is correct for your car. If you are a beginner you will have to work very methodically and there are no real short cuts to success. For example, there will not be sufficient room to allow the new wiring to be installed as the old wiring is disconnected. This idea together with any method of labelling wires and components could also fail if the colours of any of the old wiring had faded sufficiently to prevent comparison with the new wiring, or if the wiring was incomplete or had previously been altered incorrectly.

A good working knowledge of the wiring colour codes which apply to your car is essential. This knowledge dispenses with the need to refer to the old harness, except to confirm that its general layout corresponds to the new one. It also rules out the need for a wiring diagram which would probably not help you unless you were willing to devote hours of meticulous effort to translating the information which it contained into a form which would be of practical value. Wiring

colours vary between motor manufacturers in different countries, and in Britain the majority of manufacturers have for many years used the Lucas colours, the major exception being Ford.

A knowledge of the wiring colours is not enough however. You will also need to be something of a contortionist in order to cope with the work behind the dash. To complete the job you will probably need to diagnose and rectify a few faults. Dare I suggest that while some of these may be of your own making, harness makers are not infallible. There may also be problems of the type which we encountered, poor earth connections etc.

If you do decide to carry out this work yourself, make the job as easy as you can. If you have little or no previous experience it is worth spending some time during the week or two before the job is done in familiarising yourself with the wiring on your car. Obtain the new harness early and get to know how it compares with the existing wiring and how it differs. Make notes and diagrams which will help you on the day and by all means study the wiring diagram too if you have the correct one for your car. When the job is to be done have a clean and tidy area in which to work with all the necessary tools close to hand together with any new grommets, battery terminals etc which you intend to fit, and an ample supply of spare terminals. This work is best done with the minimum of interruptions so that by working methodically through the electrical system nothing is left undone. □

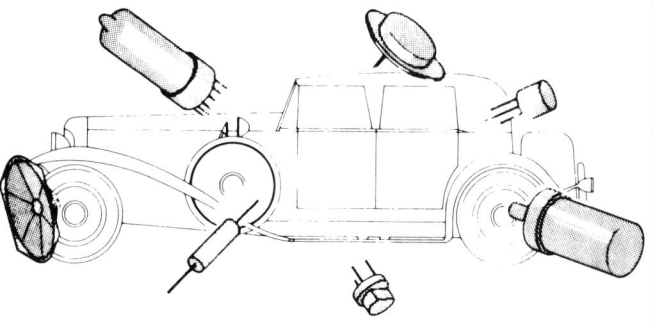

From about 1960 the number of cars which were wired for negative earthing, and which were fitted with alternators, increased rapidly, and accessory manufacturers have since produced an enormous number and variety of negative earthed 'extras' for do-it-yourself fitting. This has left owners of earlier post-war cars at a disadvantage as their cars were wired for positive earthing which does not suit the majority of modern accessories. It is not much consolation to these owners to learn that many pre-war cars were negative earthing and that the opposite polarity was adopted in order to bring about an improved electrical performance. It is very often because an owner wishes to fit a modern radio/cassette player into his car that he wants to change the polarity to negative earth, and we at *Practical Classics* have received a number of enquiries recently, as to whether this can be done and how to do it.

The short answer is yes, it can be done, few of a car's components are affected by a polarity change but it is essential to know which components are polarity sensitive and to take the necessary action.

To find out what is involved I turned again to DAK Autos at Luton where this job is in frequent demand and watched a P4 Rover being converted from positive earth to negative earth.

A brief outline of what is involved in changing the polarity might be helpful before I describe the job in detail. Put simply, polarity can be taken to mean the direction in which electricity 'travels' around a circuit. Thus reversing the polarity merely reverses that direction (as a definition this will be considered nonsense by anyone with a more advanced knowledge of electricity, but as an illustration it will serve our purpose). The electrical components of a car can be divided into groups according to how they will be affected by a polarity change. Many components will be completely unaffected (for example light bulbs, switches, solenoids), whilst some types of electric motors will simply run backwards and this can be overcome by swapping the leads to these motors. Other components, particularly those incorporating electronic parts, will not accept a change of

Old terminals are quite likely to have corroded and this is a good time to change to new clamp type terminals.

think negative

On some cars it will be necessary to extend the battery leads and swap or change the terminals when the battery is turned around. Assistant Photographer Chris Graham demonstrates that this is not the case on our Sunbeam Rapier which has long battery leads and the recommended clamp-type terminals.

polarity and cannot be modified to do so. The final group of components can have their polarity changed to suit the new arrangement, for example the dynamo, certain electric motors, and radios etc which incorporate a polarity switch.

A word of warning here; it is not possible in this short article to cover every type of component which will require attention during a polarity change, although I believe that the description which follows will be more than adequate for the majority of owners. If you are unsure about the polarity requirement of a particular component it is advisable to leave that component disconnected (with its leads taped for insulation) when you reconnect the battery and then seek further help. Do not run a dynamo with any of its leads disconnected, and do not run an alternator unless all of its leads are connected or all the leads disconnected. Bear in mind that if you are not sure which type of motor drives your windscreen wipers, heater fan etc, you can check the direction of rotation of the

John Williams looks at the mysteries of changing polarity.

motor before disconnecting the battery at the beginning of the polarity change operation and check it again briefly after the battery has been reconnected towards the end of the job. A brief test such as this should be enough to indicate whether the motor is running backwards without harming the motor, and if this is the case it can often be put right by swapping the connections to that motor.

the battery

The first step is to disconnect the battery so that it can be turned the opposite way round. This will usually entail replacing the battery leads with longer ones or at least extending the existing leads. If inverted cup type terminals are fitted (the ones which are attached to the battery posts with self-tapping screws) these will have to be swapped over as their internal diameters will be slightly different. On the other hand this type of terminal is likely to have become worn, with larger and larger self-tapping screws needed to secure it so this is a good time to change to the clamp type terminals. Leave the battery disconnected and move on to the other components.

instruments and radio

Instruments which may require attention are the rev counter, ammeter and clock. If your clock bears the information (on the back) "diode fitted, positive (+) earth only" its modification to negative earthing is not a do-it-yourself job (unless you have special skills in electrical matters). Some clocks are reversible but Derek Ramsbottom of DAK Autos recommends that you fit a quartz clock anyway.

The leads to the ammeter will have to be swapped over from the negative terminal to positive terminal and vice-versa and there may be up to three wires on each terminal depending on the vehicle.

The ammeter (at which the screwdriver is pointing in this P4 Rover) may have up to three negative and three positive leads and these are swapped to the opposite terminal.

Electric rev counters can be a problem. Some have a polarity changing switch built in and you are lucky if you have one of these. Some rev counters can be changed to the opposite polarity but bear no evidence that this is so (your local auto-electrician may be able to advise you about this), and others cannot be changed anyway so a replacement will have to be found.

A change in the radio equipment or the addition of a cassette or cartridge player is quite often the reason for wanting to change a car's polarity, but sometimes the change is made in order to accomodate an alternator instead of a dynamo (although I should point out that positive earth alternators are available). With many of the older valve radios polarity does not matter (if in doubt consult experts such as The Vintage Wireless Company, telephone Bristol 565472) but with transistorised equipment it is another matter. Some will have a switch for changing the polarity, others will not and few of these can have their polarity changed, even by an electrician.

ignition and ancillary equipment

Older coils are unpredictable as regards their performance following a polarity change. DAK Autos recommend that a good modern coil be fitted.

DAK Autos recommend that a good modern ignition coil is fitted. This will be marked + and — rather than SW and CB. Normally it is necessary only to swap the leads to the SW and CB terminals, but some coils are polarity sensitive and you may have no means of finding out whether yours is until you see how the car runs after the change.

Most alternators are suitable for negative earthing (the ACR models for example) but there are positive earthing models such as the Lucas 10 and 11 AC and these particular ones have a separate control unit, warning light unit and field relay.

Early types of fuel pump are not polarity sensitive but later types are and should have a label indicating which polarity is required. Do not try to modify the pump to suit the new polarity but buy a suitable pump instead.

Windscreen washer motors fall into two groups. Those which have two terminals

If your fuel pump is polarity sensitive do not try to modify it but buy a new one of the correct polarity.

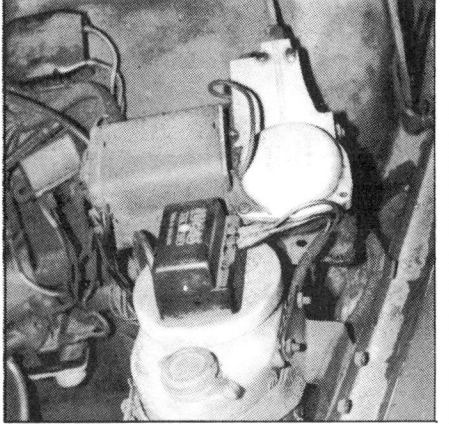

This is a "two terminal" windscreen washer on which the leads are swapped. The alteration to the wiper motor wiring (if any) can be discovered as indicated in the text.

should have the leads swapped between the terminals. On those with three terminals you should identify the terminal which has a lead to earth, leave that one alone and swap the other two.

There are two types of wiper motor, the wound field motor and the permanent magnet motor. On wound field motors a change of polarity will make no difference but on a permanent magnet motor a change may cause the motor to run backwards, adversely affecting the operation of the parking mechanism and causing more rapid wear in bearings which were designed to cope with thrust in the opposite direction. It is probably best to ascertain the direction in which the motor rotates before the battery is disconnected at the beginning of this job as suggested earlier, and swap the connections to the motor if a further check at the end of the job shows that the motor is running backwards.

Most heater fan motors have two black wires which require no change, otherwise the connections may need to be swapped and you will be aware of this if the fan sucks air out of the car instead of blowing air in.

electronic devices

In general, electronic equipment is intended for negative earthing and cannot be changed to the opposite polarity.

dynamo and regulator

The next job is to change the polarity of the dynamo. This is done by disconnecting the leads from the terminals marked 'F' and 'D' on the voltage or current regulator. The 'D'

This heater fan unit has two black wires and should therefore require no change.

lead should remain disconnected and not touching anything.

Now the battery is reconnected, the live (positive) lead being connected first. Touch the earth lead to the battery terminal by way of a test before connecting up. If the car door is open and the interior light is therefore 'on' there may be a very small spark, but any large flash indicates that there is a fault and you should check all connections before trying again.

Return to the regulator and stroke the field terminal wire (F lead) across the main battery feed terminal of the regulator about five times at one second intervals producing a spark each time. This changes the polarity of the dynamo. Note that the main battery feed terminal will be marked 'A' on (two bobbin) voltage regulators or 'B' on (three bobbin) current regulators. Having done this, reconnect the leads to the regulator terminals and it is time to start the engine.

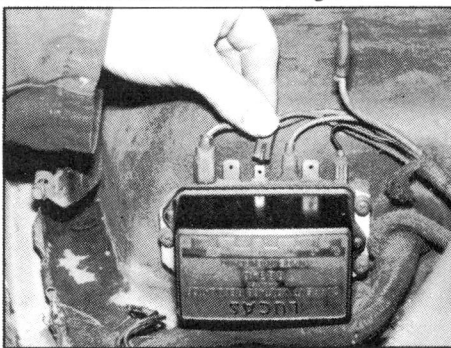

Here the lead from the 'D' terminal has been removed and is pointing upwards out of the way and the 'F' lead is just being disconnected.

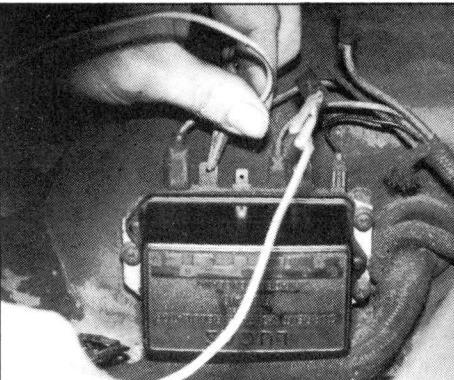

Now a short extension lead has been attached to the 'F' lead and the other end of the extension lead is being stroked against the main battery feed terminal to change the polarity of the dynamo as described in the text.

Watch the ammeter as you start the engine; it should show the usual discharge reading during starting followed by a charge reading when the engine is run quickly.

Finally, tackle this job one step at a time so that you do not forget anything, and try to arrange that you will not be distracted while the work is in progress. Take every opportunity to check the condition of wiring and to ensure that all connections are clean and tight. Loose or dirty connections reduce electrical efficiency, and increase resistance causing a build-up of heat with a consequent risk of fire. □

GOODBYE DC...HELLO AC!

Not a flat battery again! It's one of the most frustrating occurrences for any motorist, and those with the older dynamo and control box charging system are the prime sufferers. But unless 'absolute original condition' is a basic principle with the classic car owner, there's no need to suffer at all; conversion from DC generator (dynamo) to alternating current (alternator) is not nearly as diffiicult as you might imagine.

Why is it that the more modern alternator is so much better at keeping the battery charged and coping with a full load of hard working electrical accessories? It's all very easy to understand, *and* without going deeply into any electrical complications. It all hinges on the different construction of the two generators. A dynamo is a wound coil of wire rotating inside solid magnets. An alternator is the other way round – a solid rotor inside wire-wound field coils. The whole point is that the alternator middle bit can spin a whole lot faster because it's solid.

Now, go back to the dynamo. The problem with it is that it doesn't start generating current until it's turning quite fast, and at engine idling speed it's doing no good at all. OK, so you've got gearing between crankshaft and dynamo because the pulleys are different sizes; why not put a big pulley on the crank and a little one on the dynamo, so it spins faster? Well, if you did that, at high engine revs, it would be *too* fast. The only other way to generate more current is to increase the size of the generator and that has its problems.

The alternator can stay small and it can spin a lot faster because of its more solidly constructed rotor. Its output is AC current and that has to be rectified before it can charge the battery, but as it's got its regulator

This is one component you won't need. Simply unbolt it from the bulkhead and, provided you have our drawing of the connections, throw it away. If you have any doubt about the wiring, make a drawing before disconnecting, so that if all else fails, you can refit it.

The first stage in the wiring is to shorten the wires to a convenient length and join together and solder the two plain brown ones. Also join together the thin brown/yellow wire and the brown/green.

A bit of modernisation that's easier than you think! By Joss Joselyn.

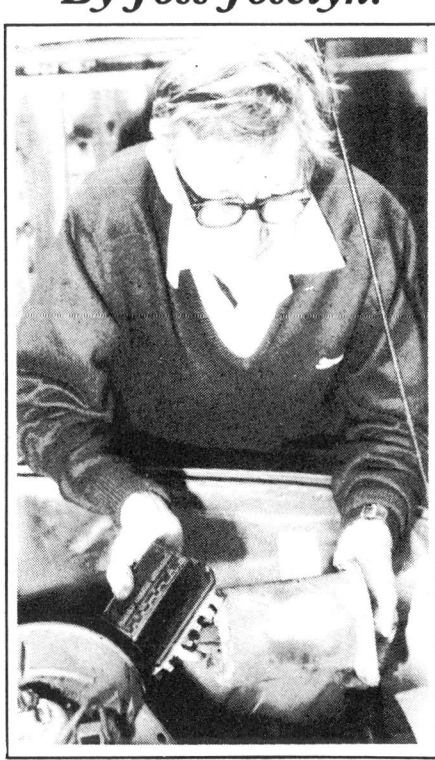

This is the sort of thing you'll be starting with – dynamo, control box and wiring. The dynamo might be a different make or a different size and the control box might vary a bit as well, but the basic wiring is not likely to be wildly different.

built in, in terms of swapping one for the other, the alternator is actually simpler.

That's quite enough theory; let's get on to the nitty gritty of conversion. First, find your alternator, and the breakers yard is probably the best bet. Go to one of the more organised places and they'll ensure you get one that works. The main other thing to look out for is

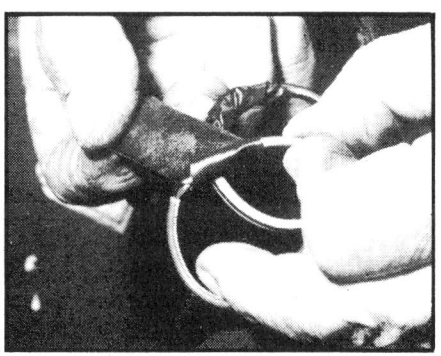

Tape the connections. Cut off the other two wires and tape the ends – that's the thick brown/yellow dynamo wire and the black Earth wire.

that you get a left-handed one if you have a left-handed dynamo, or vice versa: you can't swop from one side to the other.

Before you start searching around for parts, etc., have a look at the existing dynamo mounting bracket. You might be lucky and find it's a dual type like the one in our photographs. If it isn't, and your car is a model that was later equipped with an alternator, you should be able to buy the relevant bracket from your dealer, or again perhaps a search round the breakers might reveal one. If both these fail, a bit of bodging at the rear end of the mounting is the answer, using a long bolt and a spacer of some kind.

For those who like to do things right and have the money, there is a Lucas conversion kit, which should give you everything you need. If you're doing it with secondhand bits, the one thing it does pay to buy is the alternator connecting plug and Lucas do a small kit for that: 54960402 is the part number to ask for.

One more thing before you start chopping wires and dumping the dynamo in the dustbin – you can only run an alternator on a negative earth system. If yours is positive earth, swap the battery leads over, changing the caps to suit different sized posts if necessary, or whatever. It's not as drastic as it sounds. You might also have to change round the connections to the ammeter; it'll read backwards otherwise, and to maintain engine efficiency, change over the coil connections too. The other items that might be affected

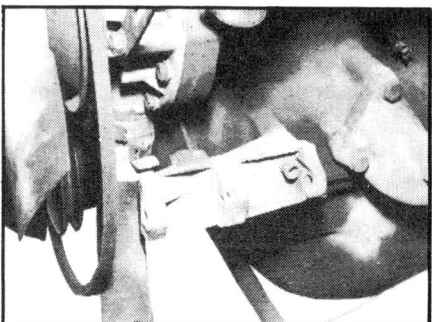

The mechanical side is easy enough on this car which has a dual-purpose dynamo/alternator mounting bracket.

If it doesn't have this sort of bracket, you can see from this juxtaposition of the two units what will be required. The front mounting stays the same, but a longer bolt and spacers will be needed at the back.

are the modern accessories like the radio or electric clock. For more about polarity changing see our February 1984 issue.

Most of the work involved is shown in our photographs and in the diagram, which may look confusing but is actually very simple when you know. Let's explain that before you start and then you can chop wires with confidence. In effect you chop off the black earth wire and the big brown/yellow dynamo wire, tape them up and forget them. The smaller brown/yellow wire on the old system goes to the ignition warning light, while the brown/green F terminal goes to the Field terminal on the dynamo. If you take the control box away and join these two last wires together, you're virtually running a wire from the alternator (when you've installed it) to the warning light. Joining together the two 'B' terminals means you're running a lead from the battery to all the external loads, like lamps, etc.

The only other wiring operation is to run a new cable from the other terminal on the alternator direct to the battery, but it should all be clear as we run through it. The conventional method is to rewire the whole shooting match, but it really isn't necessary. The way we've described is a lot simpler.

Start by disconnecting the battery and then remove the control box, pulling all the wires off as you do so. Cut all the terminals off the ends and, referring to the diagram and the photographs, complete the wiring.

The two pairs of wires that are joined together will be much safer if they are soldered. Bare a good amount of wire, twist them together and then tin them with solder so they won't pull apart. Then wrap them tidily and thoroughly with insulation tape.

If there's only a single wire on the battery terminal, you can cut that off and insulate the end.

Turn now to the dynamo. Pull off the connections, and the thick brown/yellow wire can be cut and taped off. The other one eventually will go onto the alternator. Now remove the dynamo.

The business of fixing brackets for the alternator has been covered earlier, so actually bolting the new unit in place shouldn't present any real problem. You could just check that the pulleys are in line, using a straightedge, but it's unlikely they won't be.

The old fanbelt may still fit. If there's not sufficient adjustment, you'll need a new one; the type recommended for the later model of your car (with alternator) should fit.

Back to the wiring. Snip the terminal off the thin dynamo wire and in its place fit the smaller of the two tags from the Lucas plug kit. Then buy a length of 65/0.3mm cable from your local auto-electrician. To one end connect the big tag from the kit and to the other a ring terminal suitable for connecting to the battery.

Assemble the alternator plug – small tag into the small slot; large one into the outer slot. The centre one stays blank; it's some-

Wires to Control Box – What to do with them

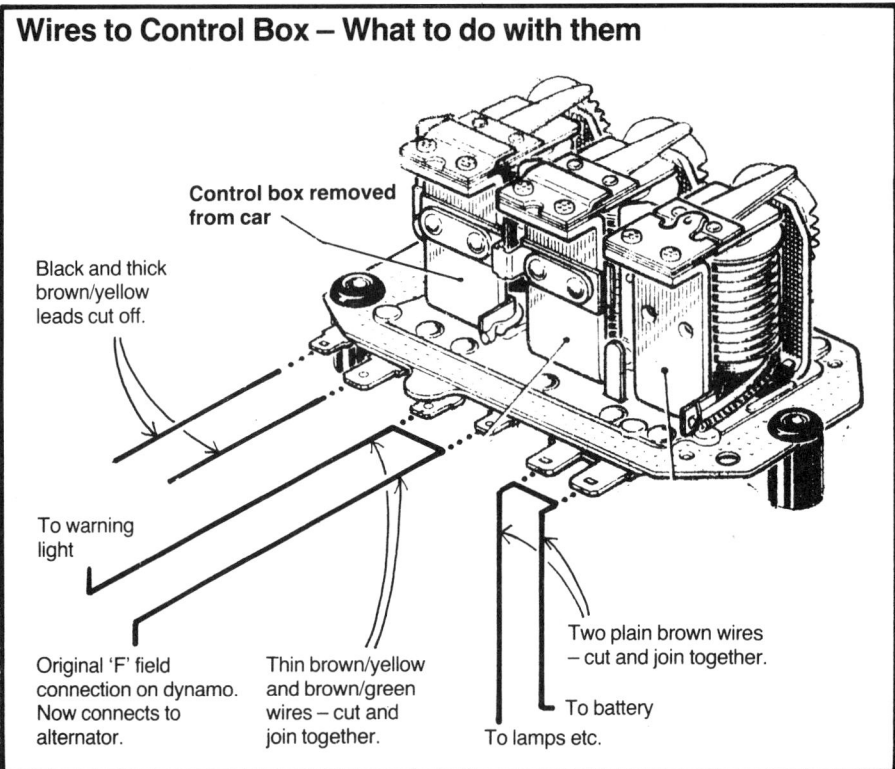

Control box removed from car

Black and thick brown/yellow leads cut off.

To warning light

Original 'F' field connection on dynamo. Now connects to alternator.

Thin brown/yellow and brown/green wires – cut and join together.

Two plain brown wires – cut and join together.

To battery

To lamps etc.

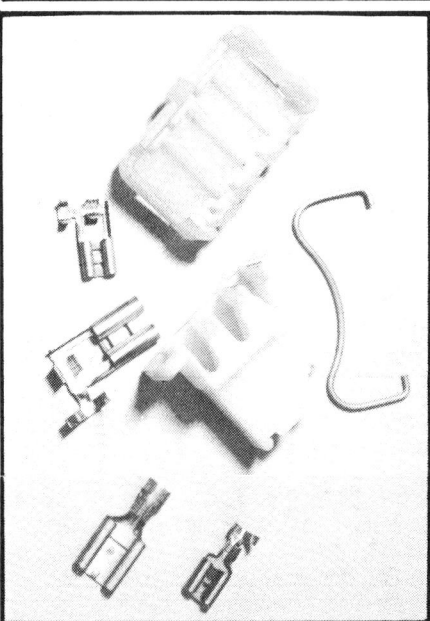

This is the Lucas alternator plug kit designed for the job. The existing tag is cut off the old thin dynamo wire, and new tag fitted. A new 65/0.3mm wire from the battery uses the other tag.

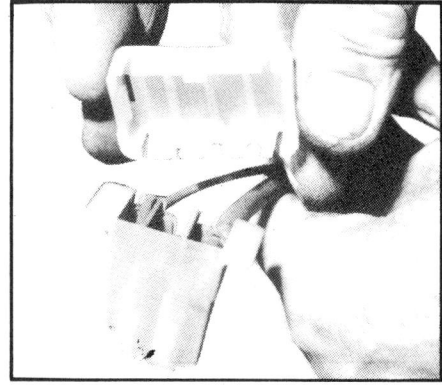

Here the plug is assembled with its two connections ready for insertion in the back of the alternator.

This is the final connection – the other end of the 65/0.3mm wire to the battery. This end has a suitable ring connector to go on the clamp bolt.

times used as a junction on some models, but normally it isn't used on a conversion job.

Push the plug into the back of the alternator, making sure it goes home correctly. The smaller connection can get distorted and pushed outside the plug.

Run the thick wire round and either connect it to the live (+) battery terminal or to the terminal at the other end of the battery cable at the starter solenoid, whichever is most convenient, keeping the lead as short as possible.

Check all the connections once more and if all is well switch on the engine. Rev it, and if the ignition light goes out, everything is working, and your flat battery problems are over.   □

ADDING EXTRA ELECTRICS

Peter Wallage guides you through the jungle of cables, switches, amps and watts

Whatever other virtues classic cars have, when it comes to electrics some of them are quite spartan in their simplicity compared with today's cars. Many fittings regarded as 'essentials' today were available only as optional extras, or from accessory makers.

In bringing a classic's electrics up to date, you're faced with the vexed question of originality. To do the job in the most electrically efficient way, the obvious starting point would be a heavy-duty alternator, a larger capacity battery and wiring laid out like that on a rally car, but it would probably lose you a lot of points from some concours judges on the grounds that it is no longer original.

However, the 1950s and 1960s were boom times for the add-on accessory people, and to my mind if an accessory is 'in period' or could have been adapted by the owners at the time that car was current, it still represents motoring of the period. And, when all's said and done, it's your car.

But we're digressing. Assuming you've decided to add some extra electrical features, the first step is to work out how much the extras are going to take out of the battery, and whether the dynamo is going to be fighting a losing battle.

The formula is very simple: amps equals watts divided by voltage. So if you want to fit a couple of 100 watt Marchal Magnum driving lights on the front then with a 12 volt system you're looking at an extra drain of 8.4 amps each. Add a heated rear window element of perhaps 60 watts, a 25 watt heater blower motor, two 60 watt headlamps, side, rear and number plate lamps at 6 watt each, a 40 or 48 watt wiper motor, a couple of dash panel lamps and the coil, all of which could be on at the same time, and you've got a load of around 500 watts, that'll take over 40 amps from the battery, so if your dynamo's, say, a Lucas C40 rated at 22 amps maximum output, or even a C40L rated at 25 amps, things are going to get steadily dimmer, and you begin to understand why 40 amp and even 60 amp alternators are available on modern cars. The thing to remember is not to have too much on at the same time.

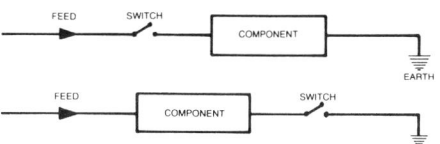

The two types of circuit with the switch before or after the component.

Lights

So, having sounded the warning, let's get down to practicalities with, first, extra driving or fog lamps at the front. To get the best out of them they must be wired with quite heavy automotive cable not, as I have seen, with a length of household flex. A catalogue from one of the cable makers will quote the maximum current the cable is designed to carry, and for driving lights you'll probably need 28/0.30 which is 28 strands of wire, each one 0.30mm diameter (In the older Imperial standard which you might find quoted in the car's handbook or workshop manual, the figure after the stroke denoted thousandth of an inch diameter, but all current cable is made to metric standards).

You've also got to look at the current rating of the switches. If you try putting 16 or so amps through a little toggle switch designed to take only 5 amps the contacts are not going to last very long. In any case, you want to keep the high-current wiring to the lamps as short as possible to avoid voltage drop and loss of efficiency, and apart from that, too many high-current cables bunched under the dash isn't the best way to avoid an electrical fire!

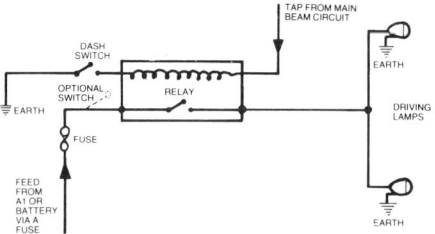

Using a relay fed from then main beam circuit to operate the driving lamps. You may feel safer using a double-pole dash switch so that the feed to the lamps is also switched, as indicated by the dotted switch on the diagram, just in case the relay fails in the closed position – though it's unlikely.

The answer is to use a relay. There seems to be a lot of mysticism surrounding relays, but they're really quite simple. All you've got inside the case is a heavy duty switch capable of carrying anything up to 18 or 20 amps (check with the specification). This switch is operated electrically by a coil which will work happily on only a few amps so the wiring from the relay to the manual switch can be much lighter than that to the lamps themselves. The coil is usually made to take quite high currents if necessary, so that the switching arrangement can be linked to another component.

For example, a legal requirement of extra driving lamps is that they shall be not used when the headlamps are on dipped beam. This can be achieved with a relay by making the main beam circuit operate the relay switch. You tap into the main beam circuit and use this as the live feed to the relay coil. The other coil terminal on the relay is taken, through the driving lamp switch, to earth. One side of the relay switch contacts is the fused live feed for the lamps, and the other side goes to the lamps and finally back to earth.

There's one golden principle to keep in mind when you're planning circuits for extra electrics. To make a complete circuit, you have the sequence of either: feed-switch-

A modern, compact fuse holder, accessible from inside the car.

component-earth, or feed-component-switch-earth. The circuit is completed by the earth connection at the battery. In the case of the driving lamps we've got things in the second order. It might all sound a bit complicated, but the diagram will help sort it out.

One of the troubles with relays is there isn't as yet a recognised standard for labelling the terminals, so each maker's labels are different. If you buy one new there shouldn't be any difficulty, but if you're looking for one at a breakers or an autojumble you might have a bit of a puzzle to sort out.

The table I've put together here isn't exhaustive, and it may not apply to *every* relay made by the following manufacturers, but it applies in most cases.

Maker	Lucas	Hella & Emu	Cartier	Mixo
Coil terminals	W1 & W2	85 & 86	1 & 2	2 & 3
Switch terminals	C1 & C2	87 & 30/51	3 & 5	1 & 4

Sometimes you might find extra switch terminals so that, for example, in the case of Windtone horns you could put 8.5 amps through each of the relay switches instead of 17 amps through only one. If in doubt, and you haven't got the maker's leaflet, try to find the circuit diagram of the car it came from and work it out from that.

On the subject of earthing, for the best efficiency I would always advise running a separate earth cable from each lamp to the chassis and not relying on the electrical contact between the lamp and its bracket, between the bracket and its mounting and possibly through rubber bushes. All these points are potential sources of trouble. The same goes for most add-on electrics. Use the same weight of cable for the earth connection as you do for the feed.

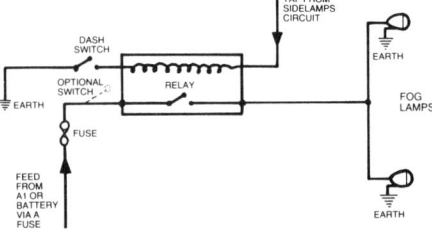

A relay in the fog lamps circuit can be used in a similar way to the in the driving lamps circuit.

You can also use a similar sort of switching control for fog lamps which are best operated independently of the headlamps but in conjunction with the sidelamps. In this case, you tap the feed for the relay coil off the sidelamp circuit. Provided you watch the current rating, the same relay can operate rear fog lamps if you fit those too. If the current is too high for the relay, use a second one.

There are a few regulations to watch when you fit extra driving and fog lamps. Extra driving lamps must be mounted only between 24″ (61cm) and 42″ (107cm) from the ground, the measurements being made from the edge of the lens, *not* the centre of the lamp. On cars registered on or before December 31, 1970, extra driving lamps must be at least 13.8″ (35 cm) apart measured from the inside edges of the lenses. On cars registered on or after January 1, 1971, the spacing is taken from the outside of the car, and the outside edge of each lens must not be more than 15.8″ (40cm) from the outside of the car.

A reversing lamp is a useful extra. Here the legal requirements are that the lamp must not have a power of more than 24 watts, which effectively limits you to a standard 21 watt bulb, but you can have two of them. They can be switched automatically by selecting reverse on the gearbox in which case you don't need a reminder warning lamp on the dash, or they can be switched manually in which case you do need a warning lamp. As you are limited to 45 watts (two 21 watt lamps and a 3 watt warning lamp) there's no need to use a relay. The total current is 3.75 amps on a 12 volt system, and it would be a very cheap and nasty switch indeed that couldn't carry at least 5 amps without burning the contacts.

You'll also want a dash warning lamp if you fit a heated rear window element, not because of any legal requirement but to avoid forgetting it and draining current unnecessarily. Some heated rear window elements have got quite a hefty wattage rating, and

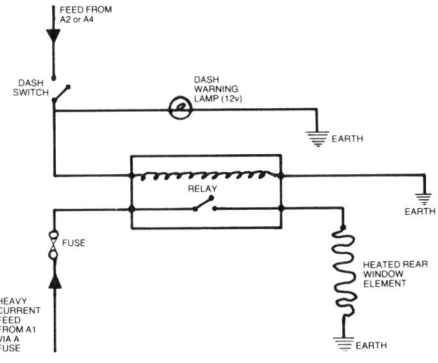

Wiring for a heated rear window. To avoid voltage drop in long cables, the heavy-current feed can, if needed, be taken from a battery in the boot, but in that case the current will not register on the ammeter.

though 5 amps through the switch will give you 60 watts, which should cover most elements, if you've got the battery in the boot it saves long runs of heavy-current-carrying cable if you tap the feed straight off the bat-

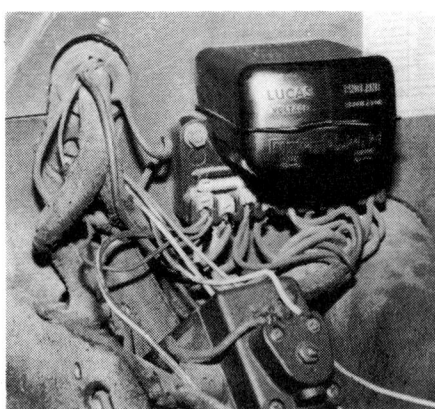

A 'classic' fuseholder, under the bonnet and not quite so neat.

tery via a fuse and use a relay in the boot which operates on an amp or so.

While we're on the subject of lamps, you might fancy fitting a map reading lamp, possibly an extra interior lamp for the rear of the car, operated either by courtesy switches on the doors or a manual switch, an under-bonnet light for dipping the oil on dark forecourts and a boot interior light to save groping about in the dark. These are all quite simple circuits, and I've covered the wiring in diagrams.

Courtesy switches are easy enough to fit in most door pillars, though you'll have to take the trim off to get to the inside, but if you decide to operate the under-bonnet and boot lights through courtesy switches make sure

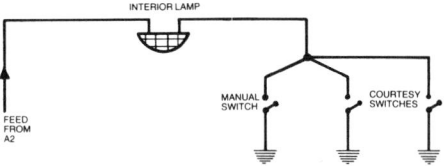

The circuit for extra interior lights.

you mount them on very firm brackets otherwise slamming the bonnet or boot lid a few times will bend the bracket and they'll stop working. My own preference is for a manual switch for these, placed where you can find it in the dark without groping. I've come across switches on the dash for them, but I prefer to have the switches under the bonnet and in the boot or you could forget and leave the lights on.

Horns

If you fancy fitting a pair of Windtone horns in place of the possibly rather feeble 'beep-beep' that came with the car you'll definitely need to operate them through a relay. Horn push contacts are seldom made to take much current, and pair of Lucas Windtones can take 17 amps compared with the 4 or 4.5 amps of an original single horn. You connect the operating coil of the relay in place of the original single horn, and take a new feed for the twin horns.

Usually you'll find that horn circuits are of the feed-component-switch-earth type to avoid carrying two wires up the steering col-

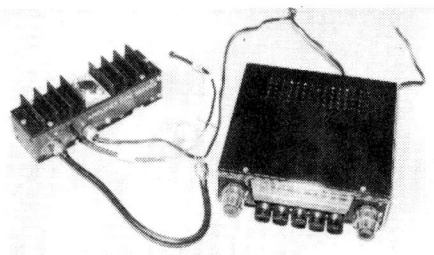

The car radio should be protected by a line fuse fitted near the unit.

umn. The earth wire comes up the column and the other side of the horn push earths to the body of the column. There's a point to watch here which can be puzzling if your new horns don't work as well as they should. Some steering columns, like that on the Morris Minor, have an earthing contact to the body with a slip ring and a brass or copper leaf contact. This can often get corroded or dirty if the car has been left standing for a year or two, so check it out and clean it.

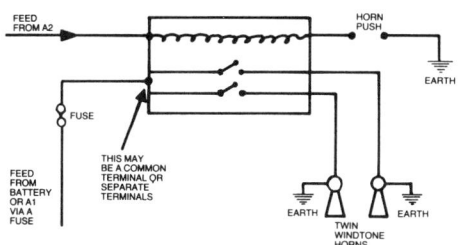

Sometimes the relay for twin horns has two internal switches to split the current. The same type of relay can be used for driving or fog lamps.

Remember that any horn will sound its best only if it's firmly mounted, so though the pair of Windtones might be mounted through a flexible steel blade, this must be fixed to a substantial part of the chassis and not to a flexible panel.

Wiper Motors

If you want to fit a wipe-wash system to the rear window, it's possible to buy a kit, but it's not too difficult to make one up from a second-hand wiper motor and screen washer. For simplicity's sake I would suggest that you go for a single-speed wiper without 'flick-wipe' facility, and though it's nice to have it self-parking, it isn't all that much trouble to look in the mirror and flick the switch at the end of the arm's stroke. It certainly simplifies the wiring as all you need is a single cable and simple single-pot switch.

When you're looking for a wiper motor for the rear window, watch for the angle of wipe. A very wide angle of sweep might be fine for the windscreen of the car it came from, but send the arm right across the bodywork when it's mounted on a rear window. You might be better off with one of the older, direct-acting wiper motors that fitted on to the screen, but sometimes it's possible with a bit of cutting and fiddling to adapt the rack drive from the later type of motor that normally sits under the bonnet. It may also be possible to alter the angle of wipe by changing the crank wheel inside the motor for one of a shorter throw,

possibly cannablised from another motor, but we're getting outside the scope of this feature.

Ammeters

If you want to fit an ammeter to see how much current all your extras are taking, it's quite a simple matter. All you do is insert the ammeter in the main feed (the main system feed, that is, not the starter feed) from the battery to the control box or fuse box. On Lucas circuits it's the one going to terminal A1. Use a good quality heavy cable if you have to rewire this feed because it carries all the load. I would suggest at least 97/0.30. If, when you've wired it in, the ammeter reads the wrong way and shows a charge when you switch something on, reverse the connections.

Current Supply

Now for where to pick up the feeds for you extras. I've dealt with picking up for the driving lamps and fog lamps, but with the others you've got to decide whether or not you want the add-ons controlled by the ignition switch. Obviously you won't want courtesy lights that don't come on till the ignition's turned on, and anything else you want to work while the engine isn't running ought to be fed independently of the ignition switch unless you've got the two-position sort of ignition switch that lets you feed the auxiliaries without sending current through the coil. If you haven't, and the engine isn't running, you're sending current through the coil all the time the ignition is on and the contact breaker points are closed. It probably won't hurt it, but there's no point in getting it hot if it can be avoided, apart from the extra drain on the battery.

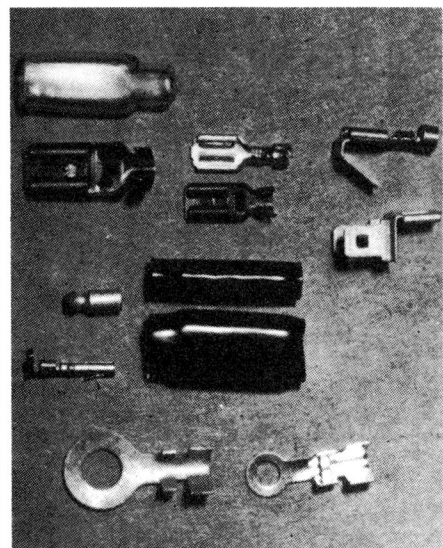

Cable connectors come in a variety of designs.

On the usual type of Lucas control box or fuse box, the feed from the battery or ammeter goes to terminal A1. This is linked by a fuse to terminal A2, and A2 feeds all the auxiliaries not controlled by the ignition switch. A1 also feeds the ignition switch

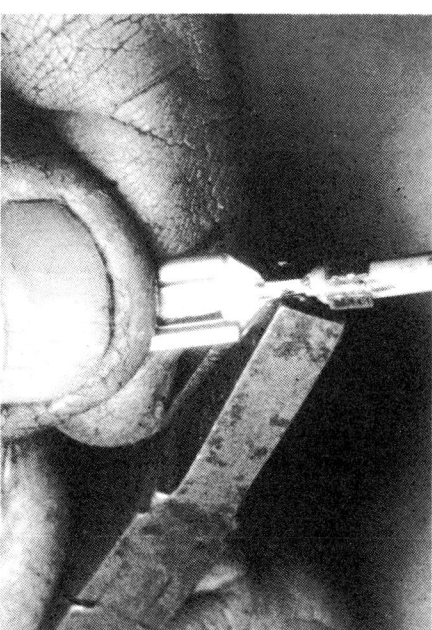

This type of connectors can be hand crimped but really needs to be soldered to ensure a good electrical contact.

direct, and from the ignition switch a feed comes back to A3. This is linked by a fuse to A4, and A4 feeds the auxiliaries controlled by the ignition switch.

To avoid bunching too many cables in one A4 terminal, there are often two of them, and there's no reason why you shouldn't add a third, or even fourth, all fed via a fuse or fuses from A3. Similarly there's no reason why you shouldn't have more than one A2 terminal provided each one is fed from A1, preferably via its own fuse. If you want to keep the area round the control box looking original, you can use a separate fuse holder hidden away somewhere (thought not so awkward that you can't find it on a dark night), or you can use line fuses in each circuit. You can get line fuse holders from almost any accessory shop.

Fuses

Choose the value of your fuses to give 100% or even 200% safety margin over your calculated steady current load, partly to allow for tolerance in the fuse making and also to allow for surge current when you first switch on, which can often be double the steady current. The highest rated normal fuse you'll get is 35 amps, and this is safe enough for most things. It will blow at the first sign of a dead short. Only if you really need to protect something, say a radio or electronic cassette system, from a high current do you need to choose a lower rated fuse, and then you usually use a line fuse fairly near to it.

Electrical Practice

There are a few practical tips that ought to be mentioned about wiring. First, please use proper terminal ends on all your cables unless they go into a socket type terminal and are held by a grub screw. The often-seen method of twisting the ends of the cables together and wrapping them round a terminal post is the mark of the bodger. You can get a kit which offers you a selection of crimp-on terminals and a tool that looks like a pair of tin snips – the blurb usually says you can also use it as an insulation stripper, a bolt cropper and several other things. Some people get on all right with them, but I don't like them. Crimping is fine if it's done on a production machine, but with hand tools I've come across too many high-resistance joins where the cable looks all right till you pull it, and then it pulls out of the crimped terminal. I much prefer the older, solder-on type of terminal.

Always use a grommet whenever you pass

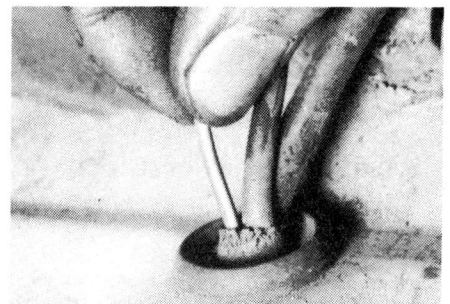

Always use a grommet when passing cables through panels – you can often 'borrow' space in an existing grommet, but don't forget to feed the cable before you solder connectors to the ends.

a cable through a hole in metal, and remember to feed it through *before* you solder the terminals on or you have to make the hole in the grommet larger than necessary. Keep your wiring neat, not like a badly constructed bird's nest, keep the runs inside the body or bonnet wherever you can, and clip them at regular intervals to the bodywork or chassis. You can buy cable clips quite cheaply at accessory shops, and you can also get pvc sleeving to protect cables which you have to run under the body or under the wings.

In future features I hope to deal from time to time with overhauling individual electrical components. □

ELECTRICAL TESTING AND TUNING

There must be many classic car enthusiasts who, like me, are fascinated by the electrical aspects of cars but are nevertheless slightly out of their depth when confronted by this traditionally mysterious subject. It is for those readers in particular who would like to do more of their own electrical work or at least tackle it more effectively, that this brief introduction to testing equipment is aimed. We shall describe what sort of equipment is available and how it can extend the scope of the DIY user.

We had hoped to sample at least two different meters from each of the five manufacturers who are known to us, but four of the firms declined to assist us. Only Gunson's had the courage to supply their products for assessment and their parcel reached us by return of post. The nice thing about it is that these products reflect credit upon the company, as we shall see.

We took the meters along to Dermody Garage at Lewisham so that we could obtain a second opinion as to their usefulness and check their accuracy against the professional Sun tuning equipment there.

The meters

We sampled three meters. They were the Sparktune at £14.38, the Autoranger at £25.19, and the Testune at £36.69. I have quoted the recommended retail prices including VAT but I understand that these meters may well be available at lower prices from some of the larger suppliers such as Halfords, Argos etc.

All three meters are supplied with clear and detailed instructions which include simple circuit diagrams illustrating the correct connections for various tests. It is essential to study the instructions carefully before proceeding, not only to avoid damaging the meter through misuse (although the Autoranger and Testune have some built-in protection against incorrect operation) but also to avoid damage to the vehicle.

Gunson's Sparktune

This is the baby of the family but it still performs a number of functions. It does not seem particularly rugged, but on the other hand my own Sparktune has survived a number of years, and obviously this type of equipment should not be transported in tool boxes, etc, where it will be subjected to knocks and vibrations – neither can you afford to drop it on the workshop floor.

The Sparktune will perform the following functions:

John Williams explores the uses of tune-up multimeters with the help of Dermody Garage.

1/ Contact breaker point condition test.
2/ Dwell angle measurement.
3/ Re-setting static ignition timing.
4/ Circuit tracing, battery condition testing, resistance measurement (leads, switches, contacts), dynamo/alternator output measurements.
5/ Measurement of resistance in suppressed leads, caps, etc.

Incidentally, dwell angle refers to the period during which the contact breaker points remain closed, and this is a much more refined method of measuring the contact breaker points gap than the use of feeler gauges. All of the meters reviewed here measure the dwell angle in degrees (it can also be expressed as a percentage), and a leaflet is supplied with each meter and provides dwell angle data for a very wide range of distributors. If, having carried out the test, you discover that the dwell angle is incorrect, you increase the points gap to reduce the dwell angle, or reduce the points gap to increase the angle, re-testing between each adjustment until the correct dwell angle is indicated.

With the Sparktune the dwell angle measurement can be carried out only when the engine is being cranked by means of the starter motor. It would be even more useful if this test could be done with the engine running as it would take more account of the effect of distributor wear at different engine speeds and an adaptor can be supplied for this purpose which costs £1.38. The Sparktunes' needle oscillates during the dwell angle test and the scale is marked in 5 degree increments, therefore absolute accuracy is not possible. However, we obtained a reading of 33-35 degrees, whereas the reading obtained on the professional Sun equipment at Dermody Garage was 32.5 degrees. This small discrepancy is hardly worth worrying about – Sparktune is still more accurate than feeler gauges.

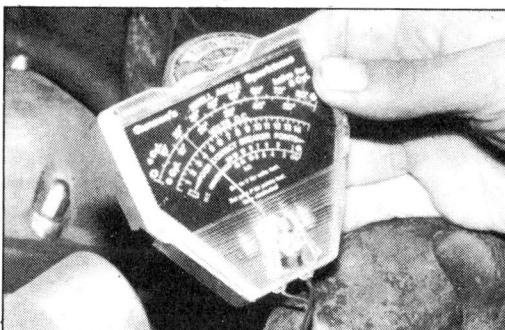

Although the baby of the range. Gunsons Sparktune is no toy. It is a useful tool, but if you require a wide range of functions and real versatility don't buy a Sparktune on impulse but save up for one of the more expensive meters.

I can vouch for the usefulness of the Sparktune as a circuit testing/tracing tool having used mine numerous times to find faulty switches or poor connections (quite often earth connections in lighting or indicator circuits). For this purpose the Sparktune is used as a simple voltmeter, and its 18 volt scale divided into ½ volt increments enables quite small variations in voltage at different parts of a circuit to be detected.

The Sparktune is supplied with over 8 feet of leads, a useful feature, but I would prefer larger crocodile clips. It will not measure the dwell angle on 6 volt cars. It is arguable that a cheap multimeter from the Far East will do everything that the Sparktune will do except measure dwell angles, but will cost less. However, my conclusions are that the Sparktune is a sufficiently accurate and reli-

Heading picture:
Our 2.4 Jaguar acted as a guinea pig when we tried out Gunsons meters, comparing their readings with those obtained on a few thousand pounds worth of Sun equipment at Dermody Garage.

Continued

able instrument to be genuinely useful. Its clear instructions will be greatly appreciated by newcomers to auto-electrical testing. It is not that expensive, and it is made by a reputable British manufacturer.

Gunsons Autoranger

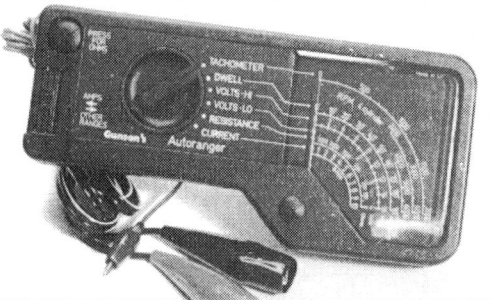

We concluded that the Autoranger is particularly suitable for owners of 4 cylinder cars with 12 volt electrical systems – which amounts to a very large market – but that for ease of use we would still prefer the Testune.

This is described as an automatic range-selection tachometer and electrical test multimeter. It works as a tachometer and dwell test meter on 12 volt systems only, but performs other electrical tests on all 6 and 12 volt vehicles. A 16 page booklet of detailed instructions is supplied.

There are high and low scales for voltage measurement (0-4v and 0-16v). For resistance measurements the scale is calibrated from zero through 5000 ohms to infinity, and this function is powered by a 1.35 volt battery which is automatically disconnected when not in use. For current measurement the scale is calibrated from 0-10 amps, with a 10 amp fuse to protect the meter. As the instructions point out, this range is insufficient to measure the full output of an alternator and any attempt to do so could result in a blown fuse *and* a damaged alternator.

The Autoranger works through three leads (about 4½ feet long, but again I would prefer larger clips) and it is a compact and handy shape. Our brief tests indicated that accuracy is perfectly satisfactory though the scales are not large and could be easier to read, and it *will* measure the dwell angle at cranking speed (enabling adjustments to the points to be carried out more quickly) or with the engine running.

It is worth repeating that the Autoranger works as a tachometer and dwell meter on 12 volt cars only. It is also worth noting that the readings on the metre are correct for 4 cylinder engines only, and have to be converted by using the charts provided on the back of the instruction book when used on 3, 5, 6 and 8 cylinder vehicles. We thought this a laborious task and felt that it could lead to mistakes being made.

In conclusion, the Autoranger is a very worthwhile tool for owners of 12 volt, 4 cylinder vehicles, but that the need to use conver-

Gunsons Testune

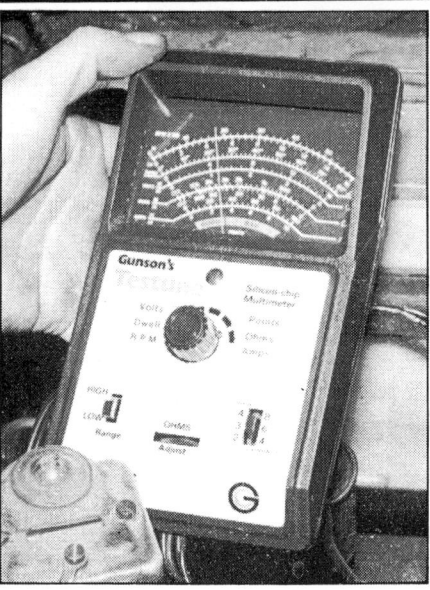

Gunsons Testune is very versatile, easy to use and has good clear scales – no doubt it is a matter of "You get what you pay for" but we think that this one would be worth saving up for.

sion tables rather detracted from its appeal in respect of other vehicles.

This is described as a silicone chip electrical test and tune-up meter for 6 or 12 volt battery, coil or electronic ignition equipped vehicles. A book of comprehensive instructions is supplied and I must say that these instructions reflect the straightforward and forthright style which I have come to associate with Gunsons, who do not make excessive claims for their products.

The Testune will test the condition of contact breaker points. It will measure dwell angles (at cranking speed or with the engine running) on a scale calibrated in 2 degree intervals from zero to 80 degrees. It will also measure resistance, current, voltage, and engine speed, and for these four functions it has high and low range scales which are used in conjunction with a selector switch. Each of the six main functions is selected by a rotating function switch and a further switch sets the meter for either 2, 3, 4, 6 or 8 cylinder (or rotor) engines. The Testune operates through two pairs of leads (over four feet long – again we would prefer larger clips, which would be easy to fit) and a copper shunt is supplied for use when measuring current between 10 and 100 amps.

The Testune measures 4½" x 8" and is therefore fairly compact, but its scales are relatively large and very clear. We found it surprisingly accurate but felt that the low range ammeter scale would be more useful if it extended to, say, 15 amps instead of 10. I also felt that some sort of securing device would have been handy to prevent the Testune from falling into the engine compartment (or wherever) whilst both hands were

busy connecting the leads. These are minor quibbles however.

Tony (of Dermody Garage) expressed some concern about the copper shunt. This is a strip of bare copper which is bolted into the circuit in which a current of 10 to 100 amps is to be measured (for example, between the altenator/generator and solenoid), and to which the blue and green leads from the Testune are then attached. It is because this naked copper shunt is intended for use in high amperage circuits and, more often than not, in those areas of a vehicle in which there is very little space to spare, that Tony was quick to point out the potential dangers. The shunt has to be very securely fitted bearing in mind that, with or without the engine running (and it will be run during this test), there must be no possibility whatsoever of the shunt working loose or otherwise managing to come into contact with adjacent body panels or other components and thus shorting out. If this happened, switching off the ignition would not help. There would be considerable damage to electrical components, probably a fire, and it could well lead to a written-off vehicle *and* personal injury.

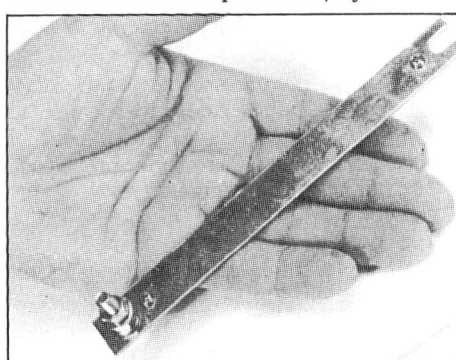

This is the copper shunt supplied with the Testune. It is a strip of bare copper about 6½" x ¾". If you use a shunt like this with any type of meter, do heed the warnings given in the body of this feature.

Having said all that, I should add (in fairness to Gunsons) that a copper shunt is a perfectly legitimate accessory – professionals have been using them for years – but we are conscious of the need to emphasise potential dangers, and especially for the benefit of our less experienced readers.

We concluded that Gunsons Testune will perform all the functions which are likely to be needed on 6 or 12 volt vehicles of up to 8 cylinders, and that it will do so with a high degree of accuracy. The scales are well calibrated and easy to read, and the high/low range selection capability is a real asset. With or without the shunt, Gunsons Testune is a very good multimeter. □

Our thanks are due to Gunsons for providing the meters, and to Tony at Dermody Garage, Lewisham, for his assistance in the preparation of this article.

CLASSICS AND IN-CAR ENTERTAINMENT

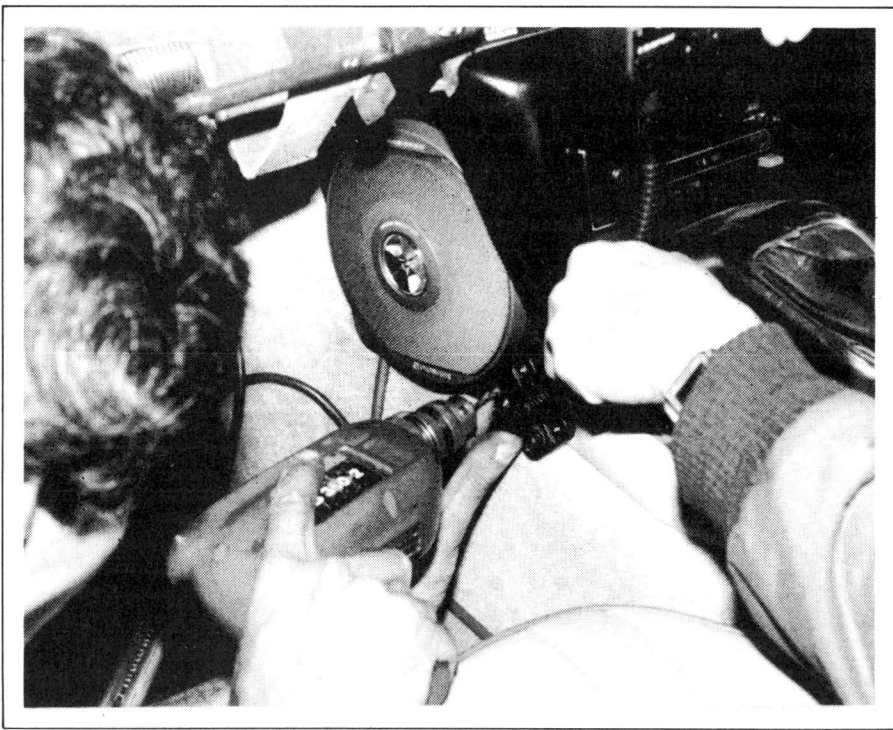

After restoration of your car has been completed, one's thoughts turn to 'in car entertainment' — a car radio, or a stereo radio cassette — on the face of it quite a simple choice really, but one fraught with problems for the owner of a classic car.

Firstly, the selection of a unit. You may choose a radio, but whether it be manual tuning or push button, the all important thing is to check the polarity of your car; most British built cars up to about 1966 were Positive earth and after that date all cars are Negative earth and you have no problems.

If your car is Positive earth, you have two safe choices; firstly you can buy a dual polarity radio or radio cassette, there being fitted with a switch or plug which is set to the correct earthing; alternatively, there are some units on the market that are so wired internally, as to take care of the polarity problem. With these you just connect the red wire from the the set to a live ignition terminal on the ignition switch or to a live terminal at the fuse box, and the black wire from the set to a good earth.

On some sets, a car radio specialist may be able to rewire the internals to give you Positive earth.

Your other choice is to re-polarize the car; this is quite straight-forward, but there are a few points to bear in mind.

To change the polarity, remove the wires from the dynamo, disconnect the battery, and using a length of wire, connect one end to the Positive post of the battery and flash it two or three times on the field terminal of the dynamo (the smaller of the terminals). You will get a small spark. Then reconnect the wires to the dynamo and reverse the connections to the coil. Reverse the connections to the battery i.e. connect the earth cable to the Negative battery post instead of the Positive, and the live cable to the Positive instead of the Negative, and re-polarizing is completed.

However, if your car has a rev counter, or a clock, disconnect these items before connecting the battery as these are polarity conscious. To get these to operate, seek the advice of the makers, i.e. Smiths or Jaeger, who should be able to rewire them, to work on Negative earth.

This also applies to some petrol pumps.

Most foreign cars of this era i.e. those with Bosch, Paris-Rhone or Ducel charging systems, are already Negative earth and require no attention.

The next step is to fit the unit, most cars will have provision for radios on the dashboard, and fixing kits for these will be obtainable from a good car radio shop. Some cars like Mini's have no provision for radio mounting, but radio mounting kits are available for these.

Wiring up the unit is again quite easy — be guided by the instructions supplied with the unit, but basically connect the live wire, which is fused to an ignition controlled power source, and the black one to a good earth; that will leave you two wires in the case of a radio, or four wires in the case of a radio cassette.

Colin Evans describes the installation of modern audio equipment in classic cars.

Connect these to the speakers making sure the wires are correctly connected. You will find one black wire which you connect to the Positive marked terminal, and a black wire with white trace to the terminal marked Negative, but again, be guided by the maker's instructions, as colours can vary.

As to the selection of speakers, most radios are supplied with a speaker which is quite suitable, but with a radio cassette speakers are purchased separately, so if your unit has an output of say 10 watts per channel, there is little point in the use of 6 watt speakers since distortion will occur. For good sound reproduction, speakers of 10 watts minimum are required.

Mounting of speakers can cause problems. With a radio you may find the manufacturers have made provision for a speaker under the dashboard, and all you need to do is bolt it in place. If not, speaker pods are available for mounting in a place of your choice.

With radio cassettes, you have two speakers to mount, these can be mounted on the rear parcel shelf, or in the doors with door mountings. Before cutting the trim, remove it from the door and holding the speaker in place, check it does not foul the window winding mechanism. If all is well mark the door panel and cut it out and fix the speaker to the door panel only (not to the metal door

Fitting car radios etc. sometimes requires a degree of ingenuity but always calls for neat workmanship and good electrical practice. The Blaupunkt radio for this E-Type Jaguar is on the end of the flexible arm. The cassette player is mounted over the transmission tunnel and the speakers are on each side of the tunnel.

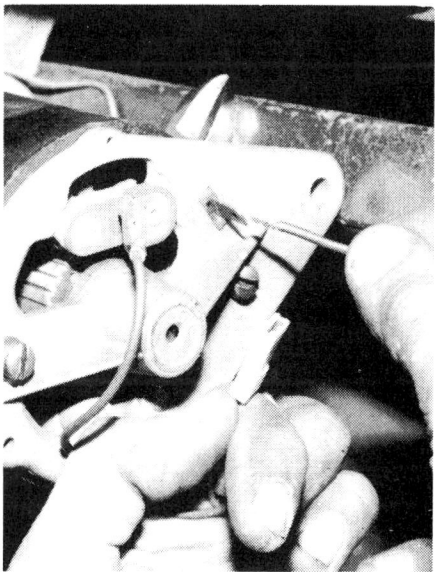

A lead from the positive post of the battery is being 'flashed' against the field terminal of the dynamo in order to re-polarise the dynamo.

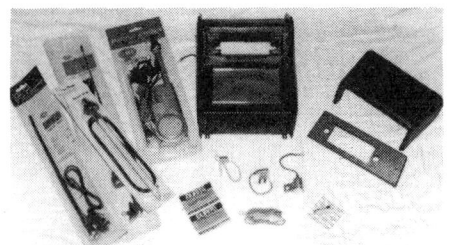

A selection of radio aerials, fitting kits and suppressors including a Hella electric aerial, a rubber FM type aerial, a radio fitting kit with built-in speaker for fitting in Minis etc, and coil and alternator suppressors.

A Pioneer speaker fitted to the rear side panel of a Ford Escort RS2000 — a neat job can be made by using narrow flush mounted speakers.

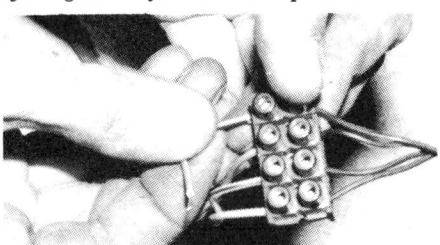

Block connectors can be used for connecting speaker wires.

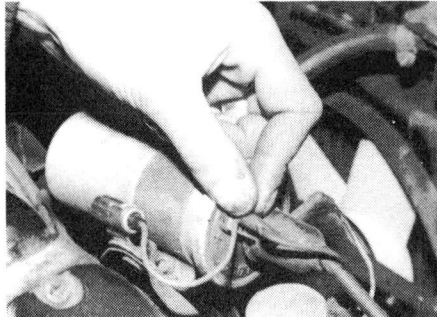

Various components will require suppressors to eliminate radio interference. These include the alternator . . . .

. . . . or the dynamo, the ignition circuit, windscreen wiper motor, indicators, clock, heater, electric fuel pump, stop lamp switch, and possibly the dashboard instrument stabiliser.

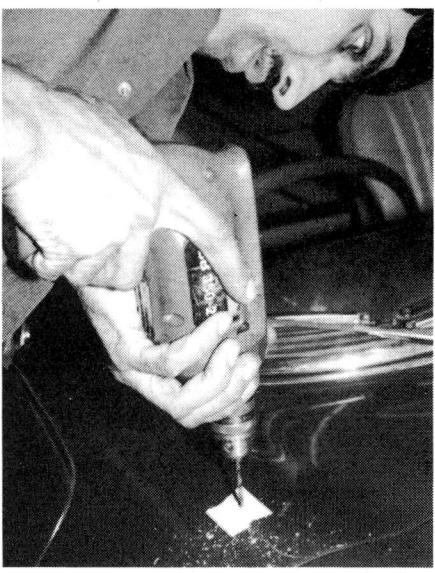

Before attempting to drill a wing or other body panel for the aerial installation use tape over the point to be drilled to prevent the drill slipping and damaging adjacent paintwork.

This is a radio and graphic unit fitted in the centre console of a Ford RS2000 demonstrating that a modern unit can be fitted neatly without looking out of place.

This is the completed installaion in the E-type Jaguar as shown also in our heading picture.

A simple way of fitting a stereo unit in a Mini; the speakers are the rear parcel shelf type but fitted upright on the dashboard.

frame as well — this will cause sound distortion).

Almost there now! Next buy a good quality aerial (not a cheap one; many a good unit has been spoilt by a cheap aerial); these are available in many types and fittings. Select a place on the wing for mounting and check you have clearance below the wing. Put some masking tape on the wing to stop the drill from slipping, drill the hole and install the aerial. Again, being guided by the suppliers instructions, plug the aerial into the back of the set, start the engine and switch on. "What a terrible row!" — which brings us nicely to suppression.

Anything which interrupts the flow of an electric current creates electrical interference; any moving part of an engine creates a frequency which the radio will pick up and which will come through the speakers as clicks, whistles or buzzing. To remove this, suppressors are fitted to various components.

Most good quality units are fitted with an automatic noise reduction circuit, which cuts out most noise problems. If your unit does not have A.N.R.C. or you still have problems, first identify the source and fit a suppressor.

Regular clicking is from the ignition circuit. Fit a 1 uf suppressor to the C.B. side of the coil; if the noise continues, fit a set of suppressed H.T. leads.

An intermittent burst of crackle that can be provoked by tapping the dashboard is the instrument stabiliser. Fit a 1 uf suppressor to the 'B' or earth terminal on the stabiliser.

A varying whine with engine speed is the generator, fit a 1 uf or 3 uf suppressor to the generator or voltage control.

Clicking when indicators are applied need a 1 uf suppressor fitted to the 'B' terminal of the flasher unit. You may find that the wiper motor, clock, heater, fuel pump or stop lamp switch need suppressing in the same way; also available are in-line chokes, which do the same type of job as a suppressor, but are fitted in-line on the offending unit.

When checking suppression, keep the bonnet closed, checking with the bonnet open will cause a massive amount of interference which will disappear when the bonnet is closed. If you still encounter suppression problems, seek the advice of a good car radio specialist.

With fibre glass cars, things are more complicated — they may need the whole under bonnet screening with tin foil to provide a good earth! In this case, again seek the advice of a good car radio specialist.

Finally, take the car to an open space, away from tall buildings etc., and tune the radio to a weak station about 200 metres on the medium wave and with a small screwdriver, tune the aerial trimming screw on the unit to give you the best reception.

Then sit back and enjoy many happy hours of music on the move. □

The writer wishes to thank Autricity Ltd of West Croydon for assistance in the preparation of this article.

There used to be an old music hall comedy sketch about motoring in which a car had broken down. The frustrated chauffeur was up to his elbows in oil and filth working on the engine, while the irate owner fumed at the delay. At one point the dialogue went something like:

Owner: "Come on, man. Come on. I said I'd got to be at the office by 9 o'clock, and what I say goes". *Chauffeur:* "Then for Gawd's sake, Guv, say 'engine'."

Had the chauffeur carried out a spot of systematic checking instead of prodding hopefully he would probably have located the trouble easily. In most cases where a classic either won't start or breaks down on the road the trouble is either ignition or fuel. As, according to the Automobile Association, there are more stoppages because of ignition troubles than fuel troubles, it's logical to check that first. We'll assume you've got petrol in the tank and that the starter turns the engine over reasonably fast.

Ignition

The whole idea of systematic checking is to eliminate each part of the system, stage by stage till you find the trouble. Start by checking that you've got high tension current going into the top of the distributor cap. Take the HT lead coming from the coil out of the centre of the distributor cap and hold it about

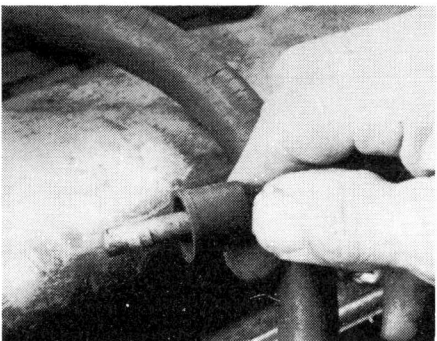

The centre HT lead from the distributor should give a good spark on a clean part of the engine.

a quarter of an inch or so away from a clean part of the engine while someone turns the engine over on the starter. You should get a good fat spark. Pick a spot away from the fuel pump and carburettor or you might get more than you bargained for.

If you get a good spark you know that the ignition switch circuit, the coil and the contact breaker points are doing their job. The

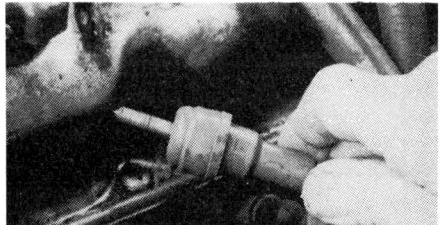

With some types of shielded plug connector a screw with its head sawn off is useful when testing for a spark.

It's A NON STARTER

Don't just prod and hope— check it systematically. By Peter Wallage.

HT current is getting to the distributor cap, so the next stage is to check that it's being distributed to the plugs. Repeat the spark check but this time put the HT lead back in the centre of the cap and try for a spark from the end of one of the plug leads. You won't get it so frequently as from the centre lead, but it should be just as good a spark.

With some types of shielded plug connector cap it's difficult to try for a spark in this way, and a useful little gadget is a bolt the same diameter as a plug connector with the head sawn off so it will fit in the end of the cap and the other end filed to a point.

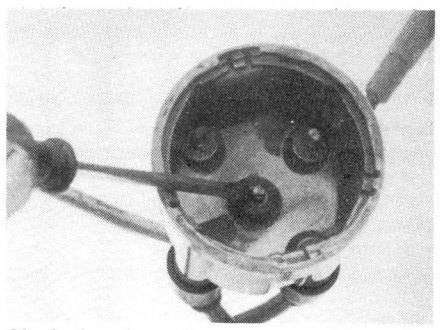

Check that the carbon contact and its spring inside the distributor cap are both in good order.

Checking for tracking inside the cap. Here there is a definite groove where, presumably, the rotor arm had hit the cap possibly because the cap was not put on straight.

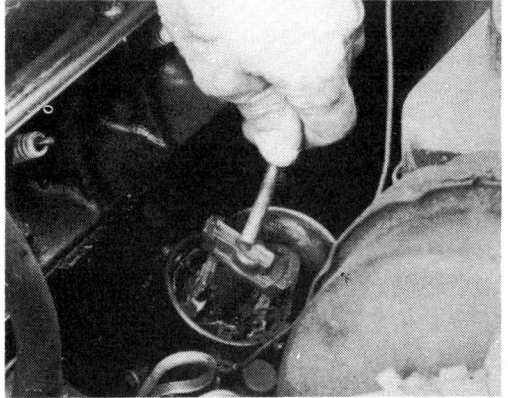

You can test for a leaking rotor arm by trying for a spark against the brass centre piece.

If you get a good spark at the ends of the plug leads it leaves only the plugs, which might be the culprit if you flooded them trying for a cold start, and if on a damp morning you got a shock from the leads it's a pretty sure sign that condensation is the trouble. But if the engine was running and stopped, and you get a spark from the plug leads, the trouble is not ignition so you can look at the fuel side.

But supposing you didn't get a spark from the leads but got one from the centre lead. Then the trouble is electrical tracking inside or on the outside of the distributor cap, the carbon brush inside the cap which makes contact with the rotor arm (or the leaf spring on the rotor arm if you've got that system), or the rotor is leaking to earth. You can check this by trying to get a spark from the end of

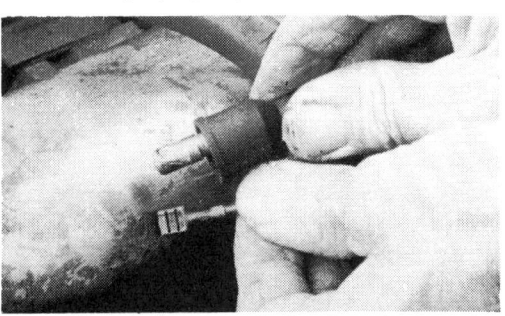

If you don't get a spark when the engine is turning over, try manually making and breaking the low tension circuit to the distributor.

An elusive fault to spot can be this low tension wire inside the distributor. Sometimes it shorts to earth.

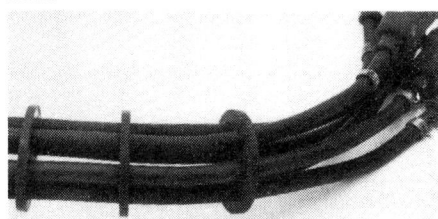

HT leads separated like this seldom give trouble, but where they run through metal tubes as, for example, on some Vauxhalls, they can short circuit one another.

the HT lead to the centre contact on the rotor arm. If you can, it's leaking.

Now let's assume you get no spark from the centre HT lead. Either the trouble is in the distributor or it is in the ignition switching or coil. Take the low tension lead, that's the one from one of the smaller terminals on the coil, off at the distributor, switch on and try for a spark from the centre HT lead while you tap the end of the low tension lead to earth. What you're doing is replacing the contact breaker points by manual switching. If you get a spark now, the trouble is in the distributor.

Check the condition of the points and their gap, and the small lead which goes from the moveable point to the low tension lead. Sometimes this shorts out against the body of the distributor and is an easy fault to miss. If the points look very black and burned the trouble could be an open-circuit condenser. The only satisfactory check is to substitute one known to be in good condition but you can get an idea by replacing the low tension lead and flicking the points open with a screwdriver. You're bound to get a small

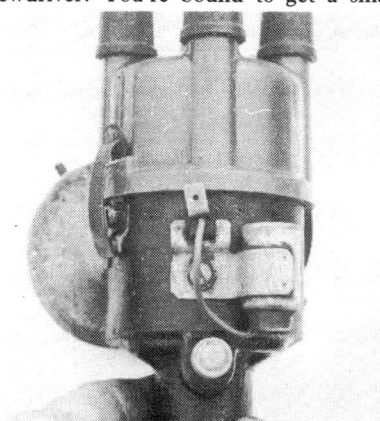

It doesn't matter whether the condenser is inside or outside the distributor so long as it goes from the low tension circuit to earth.

spark as they open, but if you get a really fat crackling spark the condenser is suspect. If you get no spark at all at the HT lead when you flick the points open, either the moveable point is shorting to earth or the condenser is shorting out to earth. If you're on the road and can't get a replacement condenser to fit inside the distributor you can connect it between the low tension terminal and earth. It will do the same job.

If you didn't get a spark from the centre HT lead when you tapped the low tension lead to earth, you've probably got trouble either with the ignition switching or leads or the coil.

To check these properly you really need a meter, but a good substitute as a circuit checker is a bulb holder and bulb with a couple of leads ending in crocodile clips.

A bulbholder, bulb and two crocodile clips make a useful circuit tester if you haven't got a meter.

Replace the low tension lead at the coil and take the other low tension lead, the one from the ignition switch, off at the coil. Clip one of your crocodile clips to earth, clip the other one to the end of the lead and switch on. The bulb should light. If it doesn't, the trouble is in the ignition switch or leads or, if you've got a ballasted ignition system, in the ballast resistor. Another possible trouble spot with a ballasted system is the second lead from this low tension terminal on the coil, the one to the starter solenoid to give an extra boost to the coil when starting. But if this is shorting to earth, the starter solenoid probably won't respond.

Assuming everything's OK at this low tension terminal, replace the lead, take the other low tension terminal off the coil, clip one crocodile lead to earth and the other to the coil terminal. If the bulb lights, the coil is OK. If it doesn't, the coil primary circuit has broken.

One other point to bear in mind if the starter is sluggish is the possibility of a poor connection in the main starter circuit. This might give so high a resistance that there is not enough voltage at the coil to let it give a

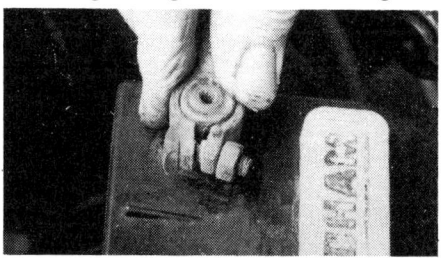

If the starter's sluggish check for tight battery connections before you blame the battery or starter.

Don't forget to check the earthing strap at the battery particularly if, as here, it is hidden away.

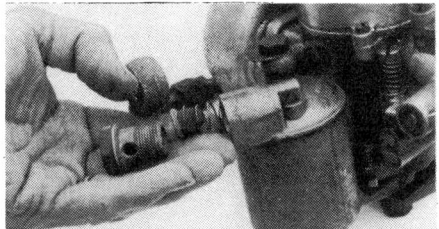

If a helmet type battery connector is corroded or just jammed, it can be freed by wrapping a rag round it and pouring some boiling water on to expand it.

good spark. Check the battery terminals, the connection of the battery earth lead to earth the terminals at the solenoid and starter and, a point often forgotten, the earthing strap from the engine to earth. This often runs from one of the bell housing bolts but might possibly be at one of the engine mountings.

Fuel

By now you should have found or eliminated any troubles in the ignition circuit so we can turn to the fuel side. An easily forgotten fault

here is that if air can't get into the petrol tank so a partial vacuum builds up and petrol won't flow out. Release the cap, and if you get a distinct hiss check that the air vent in the cap, or the breather pipe on the tank, is free.

Assuming they are, check that petrol is getting to the carburettor by loosening the feed pipe at the carb and operating the pump. If fuel comes through OK check any filter at the union, and then the jets on a fixed jet carb or, on a variable jet carb such as the SU or Stromberg, check that the piston in the dashpot isn't sticking. If the trouble is only with cold starting check that the choke is operating properly, and if it is hot starting check that the choke is coming off properly and that the float chamber isn't flooding or that the level isn't too high.

If you don't get fuel at the feed pipe start checking back along the line towards the tank. Check the feed line for blockage as far back as the pump by unbolting both ends and blowing through it, then check the filter in the pump. In a mechanical AC type pump this is under the domed top cover, and in an SU it is under a screwed plug, usually labelled.

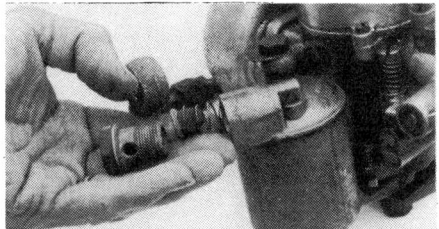

Check the filter at the carburettor and at the fuel pump for blockage.

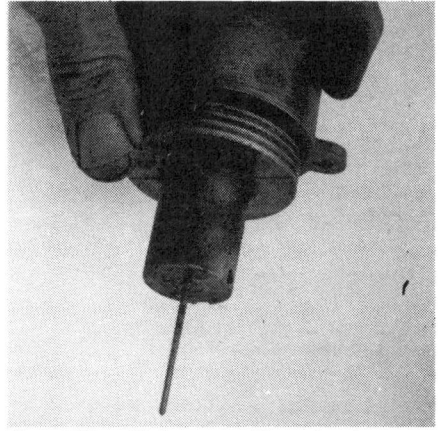

If an SU carb's been standing, the piston can stick in the dashpot and give intermittent starting troubles.

If there's no petrol in the pump, you've probably got a blockage in the pipe from the tank, or a broken pipe, or the pump isn't working. This could be because one or both of the non return valves in the pump is stuck, which could also be the cause if there is petrol but it isn't being pumped up to the carb. Or with an electrical pump it is that it isn't getting current or that the points are giving trouble. You can check the electric lead with your lamp and crocodile clips. If that's OK, disconnect the feed pipe from the tank and switch on. A pump that's operating properly should tick like mad. Sometimes, but by no means always, giving the plastic top of an SU pump a thump will start it working again.

Once you've got the sequence of systematic checking clear in your mind it takes less time to do than to describe. Checking the ignition and fuel system like this won't locate any mechanical problems that are stopping the engine from starting, but if the AA figures are right it will find the trouble in something like 80% of cases. □

"Just as I thought! You're not having trouble with the gearbox at all — you're keeping another woman in the garage!"

GENERATOR FAULT TRACING

A car that won't start in the morning, the need to keep on using a trickle charger overnight, or even the time when you have to be rescued with jump leads or a spare battery — all these things point to an ageing and tired power source. But, to quote the words of the old song, 'It ain't necessarily so!' The villain could just as easily be the dynamo or something else in the charging circuit. But what? How to track down the culprit is what this article is all about.

There are all sorts of short cuts and checks to discover what's wrong, but there's also an established test procedure, advocated by Lucas; and they of all people ought to know. Eliminating the battery from your enquiries is the first step and what you don't do is go out and waste money on buying a new one. The first test you can make is simply using an hydrometer to measure the specific gravity (SG) of the battery electrolyte.

Generally, in this country, we use a mean temperature of 60 deg. F. as a datum and, because SG is affected by the ambient temperature at which it is measured, adjust it up or down from there. At 60 deg. F (15 deg. C) a fully charged battery should produce a hydrometer reading of between 1.270 and 1.290. Around three quarters charged, the reading should be between 1.230 and 1.250 and with a totally discharged battery, the figures will be 1.110 to 1.130. In winter, with the temperature around 10 deg. F lower (50

deg. F.), reduce the figure by 0.004. In summer, for every 10 deg. F above 60 deg. F., add 0.004.

Carry out hydrometer test readings on every cell and they should all be around the same level. If they vary by more than 0.040 between highest and lowest, the battery is probably ailing.

Don't take hydrometer readings immediately after topping up the electrolyte level; they could be distorted. Take the car out for a run first, so that gassing mixes it up thoroughly; then do your measuring.

Another check that's advisable before even contemplating a new battery is a heavy discharge test, and for this you'll have to see your local garage or auto-electrician. As a result, you'll know for certain.

The next test is obvious and very simple — fanbelt tension. You don't always get the charging light showing, and a slipping belt doesn't always show up as that familiar shriek from the front end. A slight, persistent, sneaky, underhand, loss of grip can sap your battery without your knowledge quite easily. A deflection of about ¼in. in the centre of the longest stretch of belt between two pulleys is about right. Look also at the condition of the belt; any signs of cracking or other damage and you'll need a new one.

Transfer your attention briefly to the dynamo itself and check the two connections, to make sure they are both clean and firmly in place. Also clean and check all the connections to the control box.

Dynamo Check

For this and for a number of subsequent checks you'll need to use a good moving coil voltmeter (0-20V). Leaving the unit in the car, disconnect the two leads and use some cable to bridge the two terminals. Then with the voltmeter between the bridged terminals and earth (Diag. 1), start the engine and gradually increase the engine speed, while watching the dial. A reading of 12V should be achieved quite easily without racing the engine and in no circumstances allow it to rise above 20V.

If the reading fails to build up as engine

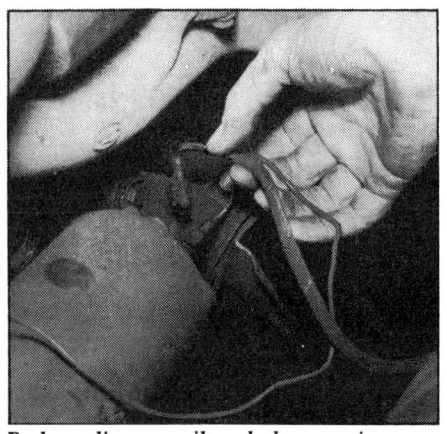

Broken, dirty, or oil soaked connections can cause problems so checking and cleaning these is one of the early stages. Look at those on the control box as well.

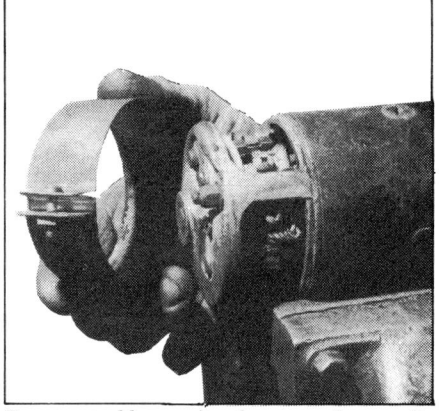

Dynamo problems quite often come down to fitting new brushes. You may be able to inspect them in situ, but you'll need to take the dynamo off to change the brushes.

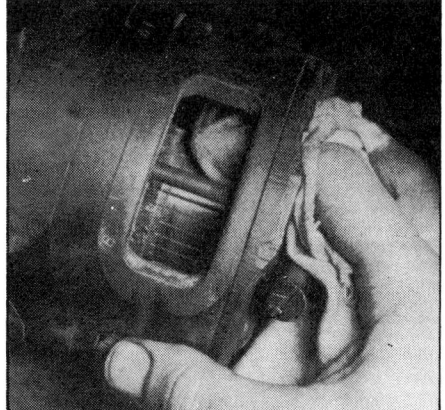

It's always worth cleaning the commutator with a petrol-soaked rag. If the commutator is scored, get it skimmed and the segments recut by an auto-electrician.

Joss Joselyn offers clear guidance on checking your dynamo/control box charging system.

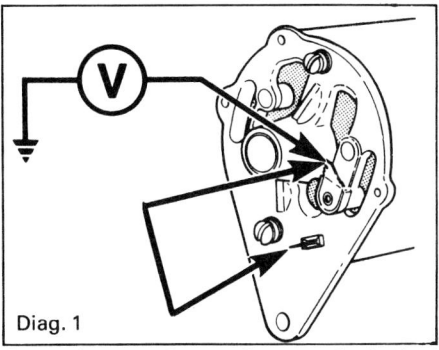

Diag. 1

Connecting a voltmeter between the bridged dynamo terminals and earth can tell you quite a lot about the state of the dynamo itself.

speed increases, there's something wrong. No voltage at all would point to a faulty armature, an unsound brushbox and terminal insulation, or faulty brushes. A reading that doesn't climb past 2-3V would indicate a break in the field coils or faulty connections.

If the dynamo is faulty, it is possible to fit new brushes and clean up the commutator. This can even be skimmed on a lathe and the segment dividers recut (something for your local auto-electrician). If a new armature is necessary or there are problems with the field coils, it's generally better to get an exchange unit.

Control Box Check

This is not something that everyone feels competent to tackle, but if you do want to have a go, we would mention that first you will need a moving coil ammeter (5-0-40A) in addition to the voltmeter. Note, also, that we tackled this subject in some detail in last month's issue. If our charge circuit check is to be complete, however, we must include it again here.

There are two main types of control box — two-bobbin and three-bobbin. In the first the functions are voltage regulator and cut-out and in the second, voltage regulator, current regulator and cut-out. Those are the three functions that can be checked.

Voltage Regulator Check

This is the left hand bobbin on both types of control box. Connect the voltmeter as shown in Diag. 2, and run the engine at about 3,000 rpm. Readings should be:-
16.0-16.5V for an RB106 (two-bobbin).
15.0-15.5V for an RB310 (three-bobbin).
14.5-15.5V for RB340 (three-bobbin).

Using a voltmeter to check the voltage regulator.

Diag. 2

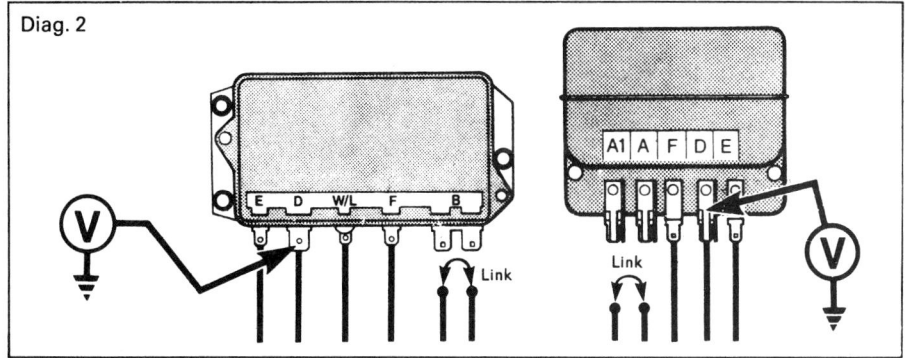

If you don't get these figures, adjust the screw and locknut on top of the bobbin. A clearance of 0.018in. should produce the right voltage which can be re-checked on the voltmeter.

If this doesn't do the trick, try the effect of a jump lead between the E (earth) terminal and a good clean earth point. If this clears the trouble, clean or make a new control box earth connection. If it doesn't, you need a new control box.

Cut-out Check

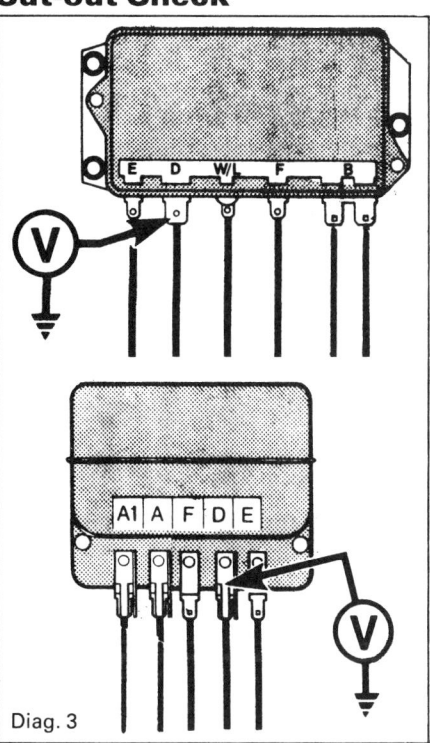

Diag. 3

Checking the voltage at which the cut-out contacts close.

This checks the voltage at which the cut-out contacts close. Connect the voltmeter between the D terminal and a good earth (Diag. 3) and switch on an electrical load (headlights or rear screen heater). Start the engine, increase speed slowly and watch the voltmeter reading. When the contacts close, the voltmeter needle should flick back between 12.7 and 13.3V. If the reading is unsatisfactory, switch off the engine and adjust the gap. Officially this should be either 0.030 or 0.040in. according to type, but if points contact is made when the armature is halfway down on

the bobbin, that's about right. If adjustment doesn't achieve the right figure, a new box is required.

Current Regulator Check

This is the centre bobbin on the three-bobbin type of control box. To make the check, the dynamo must produce maximum output irrespective of the state of charge of the battery, and this means putting the voltage regulator out of action by clamping the contacts together; a small crocodile clip will do. The ammeter is connected between the linked 'B' leads and the 'B' terminal on the box (Diag.

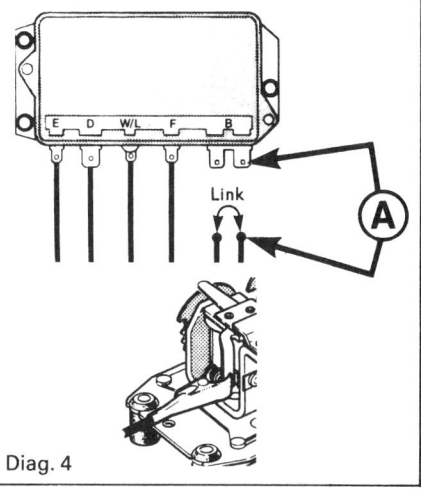

Diag. 4

Putting the voltage regulator out of action by clipping the contacts and running the engine to show dynamo maximum output on the ammeter. Do this with the headlamps on.

4), and the engine run at 3,000 rpm with the headlamps switched on. The dynamo's rated output should be shown on the ammeter (pick yours out of the table below).

Associated Dynamo	Nominal Setting
C40/1 (Fan 4½″ or	
114.3mm) dia.	20A
C40/1 (Fan 5″ or	
127mm) dia.	22A
C40A	10.5A
C40L	25A
C42	30A
C47	30A
C48	35A
C45PV/6	25A
C40T	22A
C40T 22762 & 22769	18A

If the amperage figure is too high or too low, adjust the setting and re-check. If the correct setting is impossible to obtain, fit a new control box.

Note that on the later RB340 control box, adjustments are made by means of toothed adjustment cams and to turn these a special tool is needed.

If you can't obtain any of these individual settings, you'll probably need a new control box, but the checks you've made also mean that you've got good grounds to go along to see your local auto-electrician just to make sure. Fitting a new box, incidentally, is not difficult; that you can certainly do yourself. □

BE A BRIGHT SPARK!

Auto-electrics is an area shrouded in mystery, but in reality a logical approach, a length of wire and a test-lamp are all that is required to trace most faults, as Peter Simpson explains.

Car electrical problems can be among the most perplexing of all for the non-specialist to solve, and many people who tackle all their own mechanical and bodywork repairs leave electrical work to the specialist. It is widely held that lots of expensive test-gear is needed, as well as a good deal of specialist knowledge, and that an honours degree in physics would be a great help

But in reality it is all a lot simpler than most people think. The car wiring loom may look complicated, but it is simply a convenient way of tying together all the different electrical circuits on a car. Virtually everyone knows that an electric circuit consists of two wires, one to take the power from the source to the 'consumer unit' (the light, motor or whatever is being operated), and the other to return power to the source. A switch can be incorporated in the circuit, to make or break it, and thus operate the consumer unit as required. Basically, a car wiring system consists of a large number of individual circuits, since there is either a metal body or a metal chassis running throughout the car, this can be used as one of the 'wires', to either feed the power to or take it back from the consumer unit, and avoid uneccesary duplication. Thus, one connection from each consumer unit and one battery connection are each connected to the car body, or 'earthed'. In the 1930s, most cars were wired with the negative side running through the body (negative earth), but around 1950 most manufacturers changed over to positive (feed) earth. From about 1968 onwards though, most manufacturers returned to negative earth, as it was thought positive earth increased body corrosion (especially to earth wire attachment points) which as we will see later is a common cause of circuit failure. Subsequent experience hasn't really proved or disproved this theory.

The earthed feed or return also makes tracing a fault in a circuit simpler, since it is not necessary always to run a test-lamp back to the battery. All that is required to trace a fault

We used our Sunbeam Rapier as the 'Guinea Pig' for this article. The electrics on this car were not touched during its rebuild as they seemed satisfactory then, but that was five years ago! Bob started by looking at the control box, since there was now a fault on the Rapiers charging system.

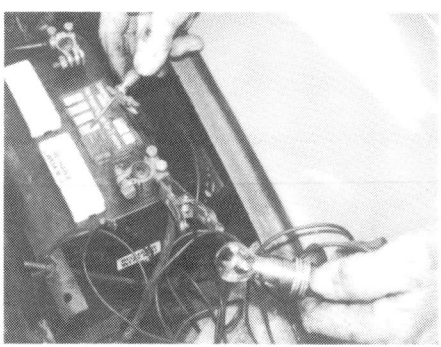

As explained in the text, the vast majority of circuit testing can be done using a 21 watt bulb and bulb holder, with leads with crocodile clips attached. As seen here, this set-up can be used to make a 'probe', by gripping a screwdriver blade in one clip.

Circuit-testing. One lead is on a good earth (the battery earthing lead in this case, but any unpainted metal surface connected to the body will do, and the other end is used to probe the circuit. Here, the fusebox is under investigation.

The fault on the charging system was that for the first 4-5 miles, the charging system worked in reverse, the ignition light glowing brighter as revs increased. Bob tracked this down to an incorrectly set cut-out in the control box.

Bob also spotted that our fanbelt was slipping on the dynamo pulley, though the tension appeared correct. He diagnosed the cause of this as a worn dynamo pulley, a new replacement is now on our shopping list!

is a test-lamp, with two leads and crocodile clips attached. Commercially-made test lamps are available, but a rear lamp holder of the snap-in type from a local scrapyard (with two wires soldered on) is just as satisfactory and much cheaper. Bob Kraft, a local auto-electrician of many years standing suggests using a bulb of about 21 watts. This will indicate whether a circuit is 'live' or 'dead' and

BE A BRIGHT SPARK!

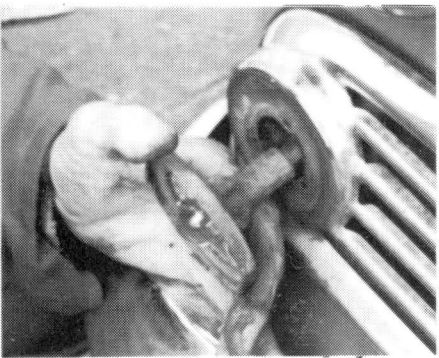

The fuse-box cover was missing, so it was hardly surprising that the connections were filthy. Bob cleaned up the contact surfaces, replaced both fuses with new ones of the correct rating, and cleaned up several of the spade terminals. He also removed a 'loose' cable running from the fusebox to nowhere, no doubt the power source for a long-removed accessory. Loose wires like this are dangerous.

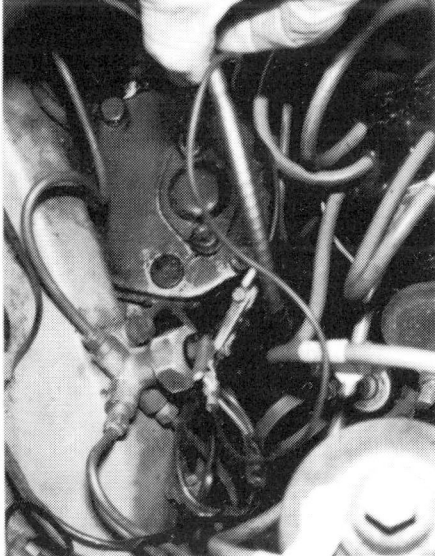

On occasions, we noticed that the flashers seemed a little slow, and the sidelamps rather dim. Inspection revealed both lamp-units to be very rusty, causing bad earth connections. Water ingress can be kept to a minimum by ensuring that foam-rubber sealing rings are correctly installed and in good condition.

(SEE BACK COVER FOR COLOUR ILLUSTRATION)
Lucas colour-coding system
Where two colours are quoted, the first is the main colour and the second a thin tracer.

Battery or solenoid switch to ammeter (if fitted)	Brown
Battery or solenoid switch to control box (dynamo)	Brown
Battery or solenoid switch to switches (bypassing control box)	Brown
Battery or solenoid to alternator	Brown
Ammeter to control box (dynamo)	Brown/White
Ammeter to switches	Brown/White
Control box to switches (dynamo)	Brown/blue
Dynamo 'D' terminal to control box	Brown/yellow
Dynamo 'F' to control box	Brown/green
Ignition switch to coil SW terminal	White
Ignition switch to ignition-controlled fuse	White
Ignition switch to petrol pump	White
Ignition switch to ignition warning light	White
Ignition switch to oil light	White
Ignition switch to accessory fuse	White/blue
Ignition switch to starter solenoid	White/red
Ignition fuse to wiper motor	Green
Ignition fuse to flasher	Green
Ignition fuse to stop-tail switch	Green. Or white on some cars
Ignition fuse to heater fan motor	Green, including Herald and Imp
Ignition fuse to reversing light switch	Green
Ignition fuse to instruments	Green
Light switch to side/tail lamps	Red
Light switch to side and tail fuse	Red
Light switch to panel light switch	Red
Panel-light switch to panel lights	Red/white
Side and tail fuse to N/S side and tail lights	Red/black or Red/brown
Side and tail fuse to O/S side and tail	Red/orange or red/brown
Side and tail fuse to side and tail lamps, both sides	BL red/green, Vauxhall red/blue
Lighting switch to dipswitch	Blue
Dipswitch to dip beam	Blue/red
Light switch to main beam	Blue/white
Wiper switch to motor (wound-field motor)	Black/green
Wiper switch to motor (permanent magnet type motor)	Blue/green, red/green or brown/green
Stop-light switch to stop lights	Green/purple (green/yellow on Fords)
Reversing light switch to reversing lights	Green/brown
Flasher unit to indicator switch	Light-green/brown
Flasher unit to indicator warning light	Light-green/purple
Indicator switch to N/S flashers	Green/red
Indicator switch to O/S flashers	Green/white
Petrol gauge to tank unit	Green/black
Temperature gauge to transmitter unit	Green/blue
Oil warning light to transmitter unit	White/brown
Ign. warning light to dynamo 'D' terminal, or alt. 'ind'	Brown/yellow
Coil to distributor	White/black
Fuse to horn	Brown (unfused) purple (fused)
Fuse to interior light	Ditto
Fuse to clock	Ditto
Horn-push to horn	Purple/black or brown/black
Horn relay to horn	Purple/yellow
Interior light to door switch	Purple/white
Earth	Black

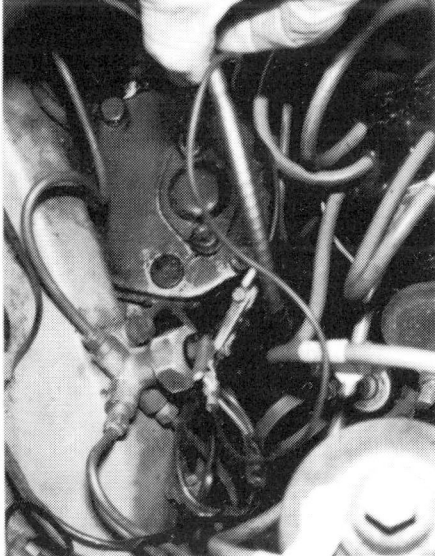

Checking the stop lamp. If the lamps do not work, connect the two switch terminals together using a short lead as shown here. If the lamps then light with the ignition on, the fault is in the switch or possibly the brake hydraulics (there could be an air bubble just under the switch).

will also show up any significant drop in amperage by the bulb being dimmer, something that a smaller lamp will not do.

To carry out the test, simply attach one lead of the test-lamp to a good earth and then, using the other, trace the circuit from the power source (the fusebox in most cases) to the item being operated, via its switch. When the point is reached where the lamp either fails to light, or is significantly dimmer than before, then the fault lies between that point and the previous point to be tested. What if you get right to the consumer unit without any change in the lamp? In that case, the fault is probably a bad earth connection from the consumer unit. When carrying out these tests, it is a good idea to put the blade of an electricians screwdriver into the clip being used to trace the circuit, as shown in the pictures. This will make access to tricky parts of the circuit easier.

A voltmeter is not the ideal tool for carrying out these tests, contrary to what some

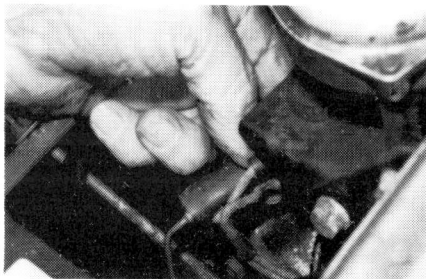

Bullet-type connectors like these can sometimes work loose, causing bad connections. Usually, they only require pushing back together. Problems of this kind are particularly prevalent in the boot, where luggage can knock against loose wires.

BE A BRIGHT SPARK!

people may tell you, since even a trickle of electricity will register as a full 12 volts. This trickle may well not be enough to operate the consumer unit, and thus the result will be misleading. Whilst we are on the subject of volts, let us be clear about the difference between volts and amps. Current, ie power is measured in amps, whereas 'pressure' is measured in volts. With the exception of the high-tension spark-producing circuits, all circuits on a 12 volt car will have a nominal voltage of twelve, but the power passing through them (the amperage) will vary a great deal. The headlamp circuit, for example, will have a much greater amperage than the sidelamp, which explains why a car will usually start if the sidelamps have been left on for a period, but may not if the headlamps are, and why operating the starter for too long can quickly flatten the battery.

To the beginner, tracing a circuit can be extremely perplexing, but simplified when all the circuits are colour-coded. Provided that the wiring conforms to original colour-codes, reference to the wiring diagram in the owners handbook or workshop manual will reveal the correct colour for the circuit under consideration. In case this is not available, we have reproduced the usual Lucas colour-coding in tabular form. Of course, there are exceptions to this, but it is the system used on the majority of UK-produced vehicles. On the subject of wiring diagrams, it is important to be aware that one might come across two types, a theoretical diagram and a physical diagram. The theoretical diagram is laid out to be easy to follow, but the position of components on it will bear no relation to their actual position, though of course they will be linked up correctly. The physical diagram does give an indication of position, but many areas on it will be cramped and hard to follow. A final thought on circuit-testing: every so often, check that the test-lamp is still working, by bridging it across the battery terminals. There are few things as frustrating as carrying out a series of tests on a complicated circuit, only to find all the work has been in vain because the lamp packed up half-way through.

Typical circuit construction and typical faults
Power supply

For now, we include under this category the battery, charging system, and everything else up to and including the fuses. As is well-known, the battery stores the electricity that drives all the cars circuits, this being generated by either a dynamo or an alternator. Alternator testing can be rather involved and require the use of special equipment, and in any case, the majority of 'our' cars are probably dynamo-equipped, so we will here concentrate on dynamos, and leave alternators for a later, special article. The dynamo will produce electricity at a rate dependent upon engine-speed, but the battery requires more power pumping in immediately after starting the engine (when the power-hungry starter has drained it) than after the car has been running a while, when only a trickle is

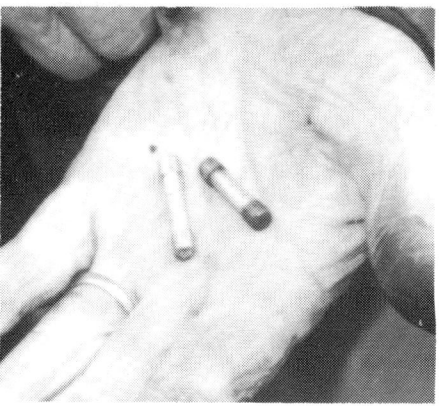

An old and a new fuse. Dirty ends can be cleaned up, but when they are really bad they should be replaced.

needed. Thus, a means of regulating the amount of electricity entering the battery is required, and this is provided by the control box or regulator, commonly known as the 'black box'. Under the cover of the black box (which should be replaced if damaged, as the regulator is very sensitive to temperature change) will be found two or three 'bobbins', consisting of copper wire wrapped around a core with an armature that can move up and down. These are cut-out devices. One is to regulate the current, as explained previously. A second is to disconnect the dynamo from the battery when the former is 'at rest'. Otherwise power from the battery would feed back to the dynamo, causing it to act like an electric motor. Where a third cut-out is fitted this (usually the centre one on a three-bobbin regulator) is the current control coil and is activated when the current passing through it exceeds a preset amount (22 amps on most British cars with the common Lucas

Another potential cause of problems on the Rapier was this badly routed battery earth cable, trapped under the battery. It should run under the carrier.

C40 dynamo). The detail of how these cut-outs perform their alloted tasks need not concern us here. Maintenance is restricted to ensuring that the contacts are clean (do not use an abrasive to clean them, the metal is very soft) and that they close or open about half-way through the armature travel. Before cleaning, measure the gap between the armature plate and core, and the armature and back-frame, and re-adjust if necessary to the correct setting after cleaning. Apart from this, adjustment in service is not normally required.

Dirty contacts are one cause of inefficient battery-charging, but by far the most common is a loose fanbelt. The usual recommendation is that it should be possible, by moderate finger pressure, to deflect the centre of

the belt's longest run between ½″ and ¾″. However, Bob recommends that the belt should be set so that it is just, but only just, possible to turn the dynamo pulley independently of the belt. This method is better, not only because it takes account of the difference in length of the 'longest run' of a fanbelt on (say) a Mini and an XJ6, but also because it shows up a worn dynamo pulley that the belt cannot grip. Other possible causes of charging system faults are worn dynamo bearings and/or brushes, poor connections, or moisture inside the dynamo. Dynamo overhaul is relatively straightforward, and we are hoping to cover it in the near future. The usual sign of a charging system that is not working properly is an ignition warning light that is slow to extinguish as engine-revs increase, or which stays on all the time, or which flickers on and off at anything above idling speed. Where an alternator is fitted, this should show a fairly hefty charge-rate immediately after starting, then settle down to show neither a significant charge nor a discharge.

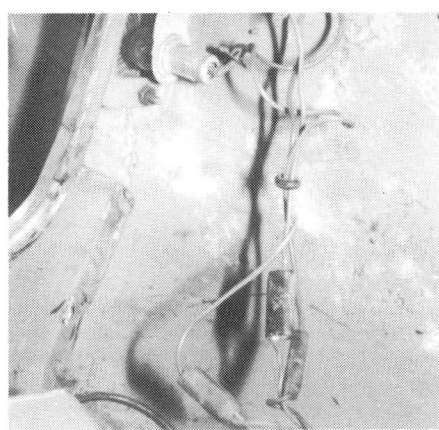

A mess like thiss, apart from being untidy and a recipe for rear light failure can be a fire risk. Imagine what would happen if a live lead came adrift and started sparking to earth in the boot. Especially if there is a can of petrol in there.....

Also included on the power-supply side is the fusebox. When two or more fuse-controlled circuits on a car fail at the same time, and they are controlled by the same fuse, it is usually the fuse or fusebox that is at fault. Do not assume though that simply replacing a defective fuse is all that is required. Why did the fuse blow? They do sometimes 'go' for no apparent reason, but it may well be that a short-circuit elsewhere has caused it, in which case the fault should be traced and rectified. If a fuse blows as soon as it is replaced, the affected circuit can be traced by disconnecting each one controlled by that fuse, in turn. Yes, you will get through a lot of fuses! Alternatively, disconnect everything, then reconnect, in turn, until the fuse blows. A fuse which goes frequently, but at no fixed intervals may be a sign of an intermittent short circuit. Fuse clips can become dirty and should be cleaned using a fine abrasive. Incidentally, it is usually the *flat* surface that is electrically important, the curved one is merely a clip. Lightly corroded fuse-ends can be cleaned up, but bad ones are best replaced with new fuses of the correct size and rating. Finally, ensure that the fusebox cover is present (they frequently are not) and *(Continued)* securely fixed in place, to minimise dirt.

Finally, consider the battery. Battery technology has improved a great deal in recent years, but batteries still have only a limited life, and 3-4 years is the most you can expect from a cheap one. Correct maintenance is of course vital, and it is important that connections are kept clean and tight. Coat them in petroleum jelly. When a starter motor refuses to turn, or turns only slowly, feel the battery connections whilst someone else operates the starter. If they get hot, they are at fault. Don't forget the body end of the earth lead, rust here is a frequent cause of poor conductivity. Where a car is used infrequently or is to be off the road for some time, remove the battery and keep it charged up on a trickle-charger. When a battery is on charge, look at each cell. If one bubbles furiously, the battery is faulty.

Starter-motor circuit

Most of 'our' cars are fitted with one of two types of starter, pre-engaged or bendix. Externally, the most obvious difference is that the solenoid is on top of the pre-engaged type.

Electrically, the circuit is simple enough, and the main starter-circuit leads can be identified by being much thicker than the others, about the size of the battery leads. A lead connects the battery to the solenoid (or switch), another (internal on a pre-engaged motor) passes the power to the motor which is earthed, the circuit being completed via the engine and an engine-to body earthing strap. The solenoid is a form of remote-controlled switch, which is needed as it is not practical to run all the power for this circuit through a dash-mounted switch. Thus, the key (or button) operated switch sends a small current down to the solenoid, operating it, completing the circuit, and operating the motor. On a pre-engaged motor, before the motor starts to turn, the solenoid engages the pinion with the flywheel ring-gear. A bendix starter relies on centrifugal force to do this once the motor starts turning.

Starter circuit faults usually take one of four forms. If the starter spins at normal speed but does not engage, there is a mechanical fault. A starter that engages but then refuses to turn may also have a mechanical problem with the pinion or ring-gear, or the battery may be partially discharged. Check it with a hydrometer, or try substituting another battery, or jump-starting. If any of these tests reveal a discharged battery, check for charging system faults and loose or dirty connections. A very common reason for a starter to turn slowly is a faulty engine to body earthing strap. Where this is at fault, choke and accelerator cables can get hot as the starter is operated, as the current tries to earth via this route. If this is not the problem then it is probably the motor itself that is at fault and it should be overhauled or replaced.

If the starter doesn't even spin, listen to see if the solenoid clicks. If it does, try bridging the solenoid terminals *carefully*, and if the circuit then operates the solenoid itself is faulty. Use a jump-lead to do this, an ordinary test-wire is not strong enough, and *take care*. The starter circuit is capable of giving severe burns. If bridging the solenoid does not work, check the lead from the solenoid to the

BE A BRIGHT SPARK!

One of the most common faults in a distributor cap is cracks, allowing the HT power to 'track' down to earth. Bob is here pointing to a favourite spot for tracking on this particular cap; between the feed cable from the coil and the cap securing clip.

motor. If this seems satisfactory the motor is at fault.

If however there is no click at the solenoid when the switch is operated, check the ignition switch and associated wiring, using a test-lamp, as explained earlier. If this does not show up any fault, check the earth and then, if there is still no joy, apply a live feed from the unearthed terminal of the battery to the solenoid operating terminal (white-red wire running to it, under Lucas scheme). If this produces action, a new solenoid is required.

On most (but not all) distributor caps, this carbon brush is supposed to be free to move up and down. Check that it does.

Ignition circuit

As most readers will be aware, the ignition electrics consist of two circuits, the low-tension circuit (LT) and the high tension (HT). The high tension, produced by the coil, is distributed in turn to each spark plug by the distributor, which also incorporates a mechanical switch, more commonly called

the contact-breaker or points. Before arriving at the coil, the low-tension current is fed through the contact-breaker points which by opening and closing at predetermined intervals only allow current to pass to the coil when a spark is required and the distributor is in the correct position to pass current to the required cylinder. To minimise arcing at the points, a condenser is fitted across them. Checking the ignition circuits is simple enough. I usually start by seeing if there is a spark at the contact-breaker points and tracing forward through the coil and HT leads to the plugs if there is, and back (first checking that the points themselves are clean, correctly set and properly fitted) if there isn't. To check for faults in the HT system, if the LT is satisfactory, disconnect the coil to distributor lead at the distributor end, hold it about 6mm from a good earth (using a pair of well-insulated pliers or a pair of the special grips available for the job), and then open and close the points with the ignition on. Make sure the points are definitely closed to start with. a good, strong spark indicates that the coil is O.K., no spark suggests the coil is defective and should be replaced. Incidentally, when replacing a coil, be certain the LT leads are fitted the right way round. The terminals can be marked in either of two ways + and —, or SW (switch) and CB (contact breaker). If replacing one type with the other, CB matches up with — on a negative earth car and SW with – on a positive earth car. If the HT output from the coil seems satisfactory, suspect the distributor cap and or rotor arm.

This month we have covered the basic principles of circuit tracing, and looked at the two most important circuits on a car, the one that makes it start, and the one that makes it go. Next month, we will move on to consider the lighting, ignition-controlled and horn circuits, and we will also look at some of the more commonly fitted electrical 'extras'.

BE A BRIGHT SPARK!

Last month we outlined the basics of how a car's electrical circuits can be checked and looked in detail at the test procedures applicable to the starter and ignition circuits. These two can be regarded as the major circuits as they are the ones which enable the car to do what it was designed for, in other words, go. This month we will look at what may be called the auxiliary circuits. These are the ones without which the car may still be driven, albeit possibly illegally, and it is faults in these that are often the most perplexing.

Due to space limitations it is not possible to cover every single fault that could occur on every single circuit on every single car. Rather, it is hoped that by using the pointers contained here combined with the testing procedure outlined last month, the novice will be able to trace and correct most faults.

Ignition-controlled lights (stop lamps, direction indicators etc)

The stop-lamp circuit consists of a switch which operates from pressure generated in the hydraulic system when the pedal is pressed, a power supply, and one or two lamps. Failure of one light is likely to be caused by an expired bulb or a poor connection, but the

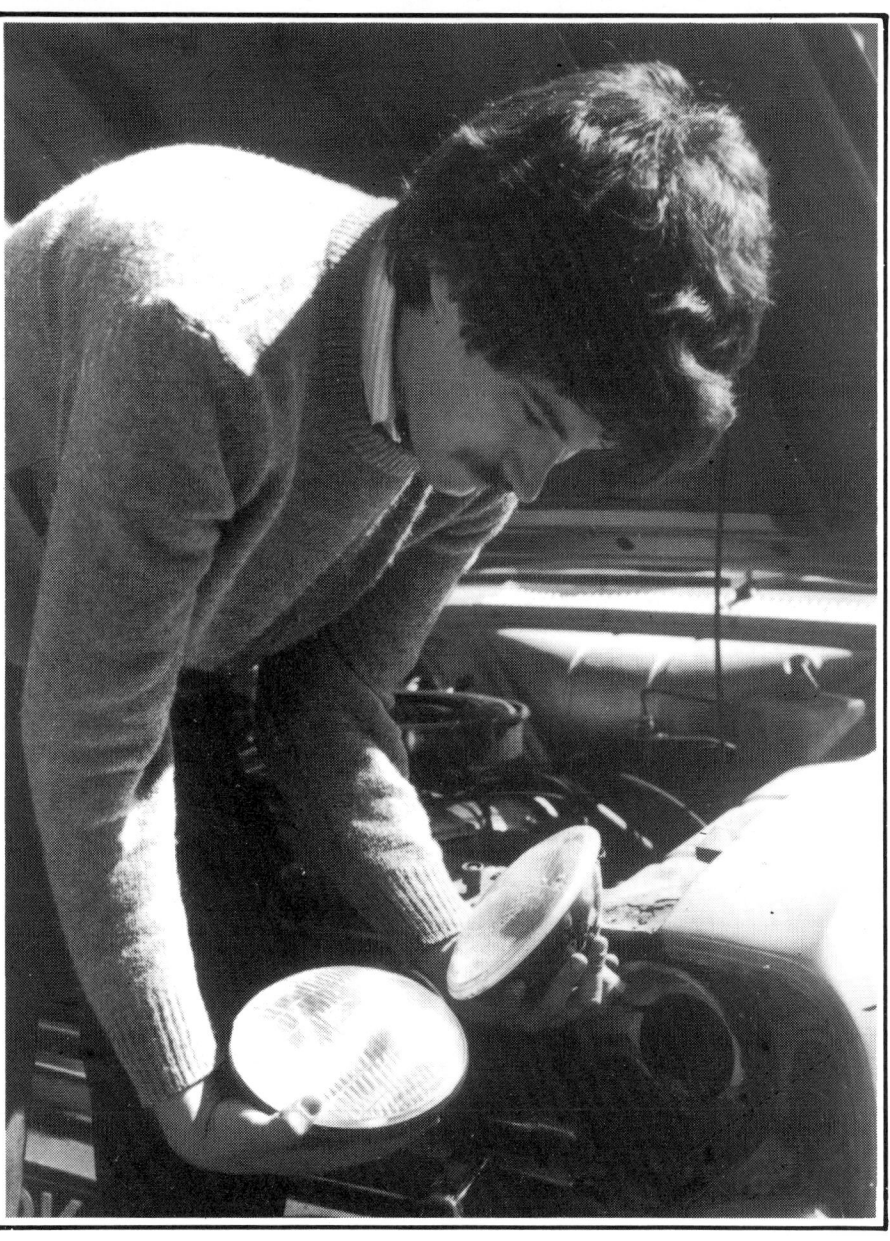

Most of 'our' cars with circular headlamps will have either Lucas sealed-beam units or Lucas '700' pre-focused headlamps. The prefocused types contain bulbs which can be replaced but sealed beam units have to be replaced as a complete unit, though this is no more expensive. Sometimes headlamps incorporate the sidelamp and our 1600E has been modified to this specification. Headlamps for some foreign cars, notably Citroens, can be horrendously expensive, so take care when fitting not to drop the unit!

---

most likely explanation of both failing, if the feed is satisfactory, is a problem with the switch. To confirm this, connect a test-lead across the switch terminals, as shown last month. If the lamps then light with the ignition on, the switch is at fault.

Some faults in the direction indicator circuit can be rather tricky to trace, since the flasher unit is designed to work with the number of bulbs in use at one time (two, or three if there is a side-repeater) so one's first assumption that a too fast or too slow flashing rate is caused by a faulty flasher unit may not be correct. In fact, if the fault seems only to be on one side the flasher unit cannot be the culprit. Check for blown bulbs, rusty holders and other causes of bad earths, and for damaged or disconnected boot-mounted lamps. Some side-repeater lamps seem especially prone to water ingress, and unfortunately once these start rusting there is little that can be done apart from replacement. On the subject of bulb replacement, sometimes a bulb will be rusted into its holder so that attemp-

Peter Simpson concludes our investigation of car electrics with a look at the auxiliary circuits.

ted removal merely breaks the glass. If this happens, try placing a cork over the broken fragments of glass still attached to the bulb base, and then turn it. If this doesn't work, the holder is probably too badly corroded to allow a proper earth connection and should be replaced.

A final word on direction indicators. From about 1965 onwards an increasing number of British cars were fitted with the Lucas type multi-function direction indicator switch. This combines the function of direction indicator, main-beam switch, headlamp flasher and horn push. When any part of these break it is usually necessary to replace the whole unit as parts are not available. New replacements can be frighteningly expensive (we were quoted £47 for a 1600E unit) so a

Stop-lamp switches either can be incorporated in the hydraulic circuit as shown last month, or operated directly off the brake pedal, as shown here on our 1600E. This type has the advantage that replacement does not involve disturbing the hydraulics.

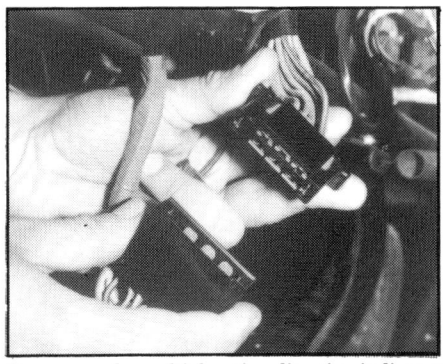

To remove the multi-function direction indicator switch, it is necessary to unplug this multi-pin connector, which will be found underneath the dash, near the steering column.

scrapyard unit would seem a better bet. After all, it is easy enough to check the condition of such a unit by inspection. Fitting is straightforward — they screw onto the steering column and the wires are connected by means of a multi-pin connector. It might be possible to make one good unit from two broken ones but don't expect a flasher switch made up in this way to last very long. Finally, these units all look identical, but in fact there are many, many variations, so make sure that the one you obtain is correct for your car.

Lighting circuits

Nine times out of ten failure of one lamp on a circuit will indicate a blown bulb. Why did the bulb blow though? It could have simply been 'old age', or it may have been caused by moisture entering the unit, or by a sudden power surge, or else by the lamp being loose, and therefore vibrating. All these should be checked if a particular lamp seems to get through more bulbs than its quota. The other usual cause of one lamp being 'out' is a bad earth. As can be seen from the accompanying typical circuit diagram, the lamps are wired in parallel; failure of one does not put them all out, therefore a total blackout is likely to be caused by either a defective feed supply or a broken switch. Of course, activating the sidelamps should, on most vehicles, also activate the panel-lamp circuit, though there is also usually another switch as well. This is after the main switch, so that the panel lights cannot be switched on without the sidelamps. Sometimes a variable resistor is incorporated so that the panel lights can be dimmed if required. Often the panel lamp switches are in rather out of the way places where they can be knocked quite easily so, if the panel lights appear not to work, check that they are switched on! Apart from that there is little to go wrong, beyond blown bulbs.

Wipers, washers and horn

On cars with a horn knob in the steering wheel centre, a wire will travel at least partway down the centre of the steering column. In time, the insulation on this can be worn through as the wheel is turned, leading to shorting-out of the cable to earth and intermittant sounding of the horn when turning corners. Intermittant sounding of the horn can also be caused by a fault in the button mechanism. Other than that, horn faults are usually caused by poor connections to the horn or horns.

Some faults on the windscreen wiper circuit can be perplexing: If the wipers appear to operate too slowly the fault may be a partially seized drive mechanism but, if the motor operates but does not move the wipers, there is obviously a fault in the drive. Self-parking wipers that do not switch off when switched off at the dashboard switch may not have a faulty dash switch. It could be that the mechanism is stiff or the motor weak. At the end of their sweep, when the wipers reach the 'park position', a switch is triggered which, if the dash-switch is in the 'off' position, switches them off. If the 'park' position is never reached, however, the wipers will not switch off. Such a fault could also be caused by the parking switch or its associated wiring.

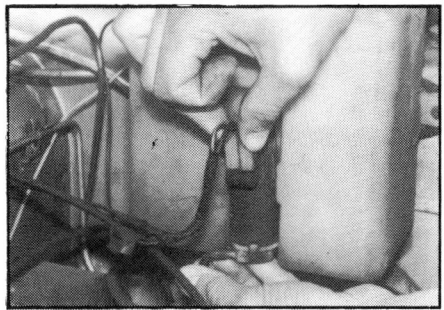

Most electric washer faults are caused by dirt clogging the pump. Check that the filter on the end of the inlet pipe is present and clear (try blowing through it or give it a quick 'blast' from a garage air line) and if a non-return valve is fitted check this is working properly, If all is well, remove the pump as shown here, and clean it.

You can check if the problem is caused by a stiff mechanism or weak motor by pressing down on the blades at the end of their run. If they then switch off, there is stiffness. Try lubricating the operating gear, and look out for obvious mechanical faults, but if no cure is found the motor is probably past its best and should be replaced by a service exchange unit. Before handing in the old unit, check on how much of the operating gear is included with the replacement — specifications tend to vary.

Most electric washer faults are caused by silt blocking the pump and stopping it working. Try blowing through to remove it; if this does not work, dismantle, clean and reassemble. Some are designed not to come apart but in my experience most can be dismantled with a little ingenuity.

Cigar lighter

A very simple circuit. Usually, power is taken from the ignition-controlled side of the ignition switch, fed to a central contact in the holder, and the main body of the holder is earthed. When the element (the bit that gets hot) is pushed in, it is held in place by a metal clip, and completes the circuit. As it gets hot, the clip expands and eventually lets go of the element, making it spring out of contact. To test, check for power at the central contact, and check the earth connection. The current pick-up point in the centre of the element can get dirty in time and prevent the circuit being complete. If all these do not reveal any problem, the element should be replaced.

Overdrive

Overdrive circuits vary in detail but all work in basically the same way. There are two switches, both of which have to be 'on' for the overdrive to engage. One, the inhibitor switch is on the gearbox and will only be 'on' when the car is in gears on which overdrive is available. The second is, of course, hand-operated by the driver. Sometimes, there is a third switch. This is normally in the 'on' position but turns off when the accelerator pedal is pressed hard to the floor, thus providing a kickdown facility. Usually, the hand-switch is relay operated, the switch operating a relay which then operates a solenoid. Checking for electrical faults should be straightforward. Failure of the solenoid to work even though current appears to be reaching it could be caused by the contacts needing adjustment, or the solenoid may need replacing. Inhibitor switches can also sometimes be troublesome. The overdrive is actually engaged by a solenoid which has two sets of windings, one to engage it and one to hold it in. Failure to engage could be caused by the former being at fault, and failure to stay in overdrive by failure of the latter.

Conclusions

Due to limited space, it has not been possible to do more than skim the surface of this fascinating subject, and I would be the first to admit that there are many areas that have not been covered, or covered in sufficient detail. To those who would like to delve further, may I recommend the purchase of Joss Joselyn and Bob Krafft's book 'Auto-Electrics; Maintenance, fault-finding and repair' (published by Newnes Technical Books). This useful volume covers every aspect of auto-electrics and is written to be understood by the layman, whilst not being superficial. Hopefully though, my attempt will persuade more people that, contrary to popular belief, most electrical problems can be sorted out at home, with a minimum of equipment and fuss. □

Thanks to Bob Krafft for his help during the preparation of this article.

Typical side/rear light circuit

Sidelamps

Earth

Battery

Switch

Number plate lamp

Tail lamps

Notice that all lamps are individually fed and earthed so that failure of one doesn't break the supply to the rest. Side and tail lamps are often used in pairs, though sometimes both lamps on one side will be protected by one fuse whereas on other vehicles the nearside front offside rear (and vice-versa) will be on one fuse.

AUTOELECTRICS

The majority of vehicle breakdowns result from electrical system faults, a known fact borne out by the findings of the motoring organisations. This leads to one simple question. Why?

The answer is equally simple. To the average motorist, car electrics are on a level with alchemy or the mysteries of creation. If the system works, fine. If it doesn't, panic; almost an automatic reaction. This isn't so hard to understand when the electrical system, with its maze of coloured wires, components and connections, is viewed as a whole. The secret lies in breaking the system down into manageable parts. Thus each circuit can be seen as having a beginning, a middle and an end. The London Underground works on the same principle and there are those who have mastered that. Therefore, by applying the analogy further, we can substitute the flow of electrical energy for the passenger and the circuit for his route. The electrical system has a number of such routes, bound together for strength and convenience into a wiring loom. The Underground works in a similar way and, to find your way about, you follow a particular colour on a map. As far as wiring is concerned, you also have colours to follow; for tube map, read wiring diagram.

Back to school

Before examining the pathways of electrical energy, confusion can be avoided by a swift examination of how electricity works. Two types of material are involved, that which conducts electricity and that which acts as an insulator. The difference lies in these materials' atomic structure. Any physics teacher worth his salt will tell you that all materials have such a structure and it consists of particles with a positive charge (protons) and particles with a negative charge (electrons). These particles are united in an atom, which always has a positively charged nucleus orbited by negatively charged electrons. The fundamental difference lies in those electrons. If you consider the atom's nucleus as a stable, inert entity, surrounded by fickle, wayward electrons, you are getting the idea. Those materials which conduct electricity hold a greater quantity of those mobile electrons, aptly called free electrons, than an insulating material. In the absence of an electrical stimulus, the electrons get their act together and pass through the material in the same direction – an electrical current.

This is a rather simplistic explanation of course, although it will serve our purpose here. The important thing to remember is that the free electrons mentioned exist, whether they are moving at random or flowing as a current. That which makes them move completes the picture. To illustrate

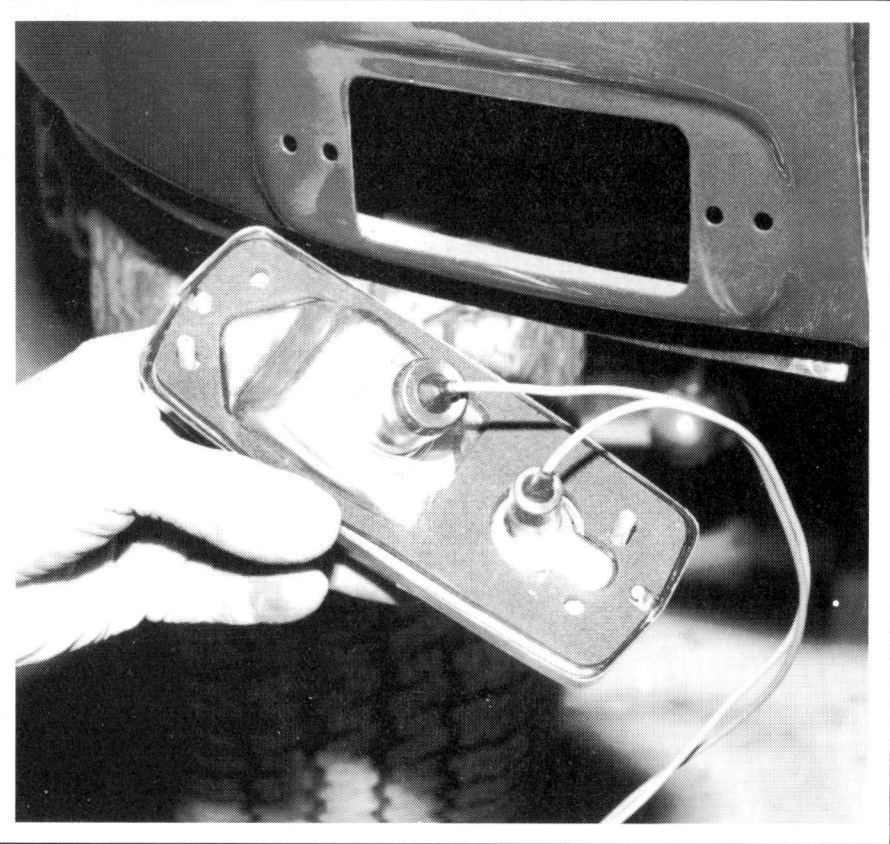

The Idiot's Guide to Autoelectrics
Dealing with car electrical systems has been known to fill the home restorer with terror.
David Hill explodes the myth.

Clean, dry and tightly connected; this car is a good starter.

AUTOELECTRICS

The rest of the system may be perfect but, if the points have closed up, there won't be a spark at the plugs.

this, we have to go back into the realms of physics. If you were to take a bar magnet, mount it on a spindle and spin it, nothing much would happen. Put a loop of wire around the magnet and you would have created a simple dynamo. As the poles of the spinning magnet pass the wire, the wires' electrons are stimulated into movement, setting up a current within the wire as mentioned previously.

The problem with this experiment is that, with the ends of the wire loop free, there is no evidence that such a current is flowing. In other words, there is no circuit. Putting a bulb across the ends of the wire would solve the problem. A simple circuit would be created, the bulb would light and everyone would be happy. There are conclusions to be drawn from this. In physically spinning the magnet you are expending energy to create electricity. In other words, you don't get something for nothing, which is the basic premise of generating electricity, from our tiny dynamo to a nuclear power station. This is where a lot of people come unstuck in the

This distributor cap has a fixed centre contact. Some have a sprung carbon contact; either way, make sure its there.

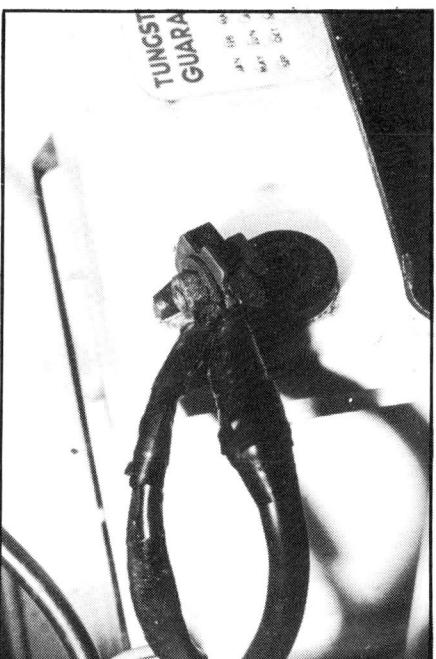

Blindingly obvious perhaps but a poor battery terminal will cause a massive power loss.

matter of batteries. If you charge your battery, whether via your car's charging system or with a mains charger, you are putting energy into the battery. The battery can store electricity but it cannot create it.

The key phrase is a simple one: potential difference. A battery has a potential difference between its terminals. Our simple dynamo has a potential difference across the ends of its wire loop. Putting a wire, bulb, or whatever across those contacts with a potential difference will set up a flow of electrons, an electromotive force (emf). Gentlemen, we have electricity.

An Ohm of your own

Like any force, this electromotive force has been provided with a system of identification. Returning to our Underground analogy, we can consider this system as follows. Suppose there are 100 passengers making a journey by tube. They represent a number of bodies (quantity), a number of passengers per train (pressure) and a number of passengers passing a single point on the journey at a given time (rate of flow). In electrical terms, the unit of quantity is the coloumb. The unit of pressure is the volt and the rate of flow is put in terms of coloumbs per second, better known as amperes, or amps. The coulomb is largely irrelevant as an isolated unit in our case, leaving volts (pressure) and amps (flow rate) as the major protagonists. Yet there is still something missing, a means of making use of this force. Without such a means of consumption, the energy would disperse itself by other means. Anyone who has dropped a spanner across the terminals of a healthy battery will know all about this. The resultant circuit would have been a very short circuit indeed and all the flashing sparks rep-

resent the violent dispersal of electrical energy in the form of heat and light.

Consequently, the electrical components of a car are designed to make use of that force, turning it into heat, light, rotary or linear motion as required. In doing so they consume current so that the current going into the component will be more than the current leaving it. The current is therefore resisted from passing freely through the circuit, hence the term 'resistance'. This value is measured in ohms. This is as far as we need go into the complex science of electricity. Our needs concern the use of an invisible force in a practical manner. Without electricity our cars wouldn't run at all, as a failure of the electrical systems will prove. Conversely, the fact that all electricity is produced and managed in the same basic manner allows us to master it and make use of it.

A ballast-resisted coil normally runs at around 7V but is allowed 12V for a better spark on starting. Using such a coil without the resistor is dangerous.

Problems, problems

Electricity is queer stuff and, in certain ways, it has a mind of its own. Unlike domestic wiring, car systems make use of a two-terminal method. The current runs from the live side of the system, through the electrical component and then to earth. Since 12 volts are not enough to cause significant damage to us if left untrammelled, the conductive steel shell of the car is used as a massive earth terminal, positive or negative, dependent on the age of the car. This is the reason why cars with non-conductive bodywork, ie. glassfibre, need separate earth returns to the chassis. At 12 volts the nature of electricity is such that it will not cope with loose or dirty contacts and circuits so compromised will fail to operate adequately. As the voltage rises, as with the 5,000 to 20,000 volts encountered in the ignition system's high tension circuit, the electrical energy behaves more like its big brother, lightning. As you will doubtless have noticed, lightning likes to go to the earth terminal. This is literally the earth, unless you happen to be in the way. In this case, you, your television aerial, your tree or your roof are the pathway to earth and the lightning

No prizes for guessing which is new. The light unit on the left caused an MoT failure.

will take it. High tension voltage behaves in exactly the same way. If your distributor cap is damp or dirty an easy path to earth is offered. The HT voltage takes this path, as opposed to going to the sparking plug and there you are, late for work.

A moment's consideration will show that, given the conditions in which they operate, it is a marvel that car electrics work as well as they do. Domestic washing machines are wired in a similar way to cars. Would you expect your Creda/Hotpoint/Hoovermatic etc. to work if you kept it outside in the rain and frost? No, neither would I. Small wonder then that the combined effects of heat, cold, damp and vibration wreak havoc with your electrons.

Sherlock Ohms

I would be the first to admit that a frigid winter's morning is not the best time to go trouble-shooting. On the other hand, necessity is the mother of invention, as will become clear. The prime requirements for starting a cold motor are a fit battery and a good spark. The presence of thick, cold oil and tight clearances, together with a cold battery's reduced output, make the latter especially important if a start is to be achieved. Around 100 revolutions of the crankshaft per minute will be needed; few engines will start with fewer. Thus, flogging a flagging system is a waste of time and you would be better employed in seeing to the ignition hardware instead. Unlatch the distributor cap and wipe off any condensation from both the inside and outside, treating the coil's nose to the same attention. Ensure that the cap's central carbon pencil or contact stud is in order and wipe off the HT leads. Dry the contact breaker points and plate and make sure that a gap exists. Replace the bits, not forgetting the rotor arm, and try again. If all else fails, remove the sparking plugs and warm them in the oven. Incidentally, I don't mean the microwave oven, which will give you as many sparks at the plugs as you want, prior to breaking. If this trick fails, beg a jump start, take the bus or read next month's instalment.

Trouble-shooting of a less desperate nature often involves the rectification of

Period electrical accessories will enhance your car. Choose carefully.

failed circuits. This involves finding out where the power is – or isn't, as the case may be. A test lamp is handy in these circumstances. This simple device may be bought although a neat one can be made using a spare warning lamp or side/tail lamp bulb holder and a pair of wires. By attaching one wire to a convenient earth point you can use its twin as a probe and, if power is present, your bulb will light. A further refinement is the addition of crocodile clips and a sewing needle that can be pushed through the insulation in mid-wire. Being conductive, the needle will allow you to search for breaks in the wire without damaging the covering unduly. Alternatively, you could be very flash and buy a test meter, which would also allow you to measure circuit continuity and the like.

While immersed in the morass, it is

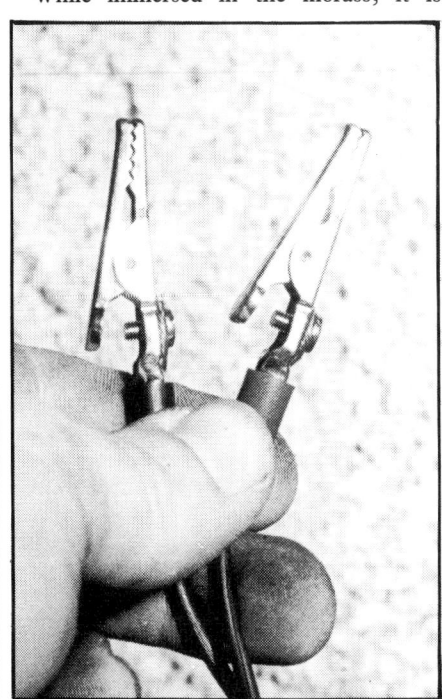

The home electrician's friend. These came from a radio spares shop and were much cheaper than a tow home.

ridiculously easy to overlook the obvious. A wire which has physically unplugged itself, or one whose terminal has fractured, invariably will wander off and hide, so check everything. A blown fuse may simply have died of old age, so it is worth sacrificing another before attempting to chase a short circuit. If you do get a short, have you worked on the car recently? Wires impaled by a self tapper or a drill tend not to work too well.

The Twilight Zone

The lighting system of a car arguably has a harder life than the rest of the electrics. Standing in the firing line, at the unprotected extremities of the car, the connections receive more than their share of adverse conditions. The consequence of this is that the lights go out, which is a distinct problem and illegal to boot. The short-term remedy for this lies in a temporary jumper cable. This, a length of wire with a crocodile clip at either end, can be carried in your toolbox for such emergencies, allowing you to carry out a proper repair at your convenience. However, it is wise to test the bulb before assuming that the wiring is at fault. A similar emergency repair concerns the tail lights. If these have failed inexplicably, remove the wires from the stop light switch and connect them together. Your temporary tail lamps may dazzle following drivers, although some cars will allow you to swap the connections and restore the tail lamps. Either way, its preferable to an improper advance from an articulated lorry, or from a traffic policeman for that matter.

Spot the distributor! Some cars, such as this Matra Bagheera, have poor access and conversion to electronic ignition has made life easier.

Full supporting programme

These bijou dodgettes will keep the wolf from the door for a time but, like preventive medicine, preventive maintenance is a good idea. The next instalment will consist of a breathless gallop through your electrical componentry to show how things work and, more importantly, why they don't. Stay tuned! □

AUTOELECTRICS

L ast month, in the first spasm of this electrical opus, I explained the basic principles of electricity. The next logical step is to examine the ways in which electricity is used in your car.

Electricity is unique in its versatility. It can provide heat, light, sound, visual information and power, as a linear or rotational force. It can be manufactured 'on site' and stored until needed, yet it has no smell, is silent, will not in itself explode or burn and in the event of its leaking away, it will leave no trace. This leads to the conclusion that those poor frogs who lost their lower limbs at the hands of Galvani and his like didn't die in vain, although that knowledge is doubtless of small comfort to them now.

However, like any other mechanism within a car, the electrical system works in a hostile environment. For all its good points, electricity is lazy stuff and would rather leak away to earth or stop at a broken contact than do its work. So, to get the best results from your electrical system you must maintain it. The snag lies in the system's complexity. The point of this feature is to explain the workings of the system and the faults which commonly afflict it. The best method of doing this is to consider the system in bite-sized pieces and, since the battery represents the heart of things, it makes a good starting point.

Assault and battery

A popular misconception is the belief that a battery can create power. This is not so; a battery is a storehouse, not a powerhouse. Last month, I used the London Underground as an analogy for the car's electrical system, with the passengers representing the free electrons whose movement constitutes an electrical current. In this context, it would be convenient to consider the battery as the city to which the passengers travel. Unfortunately matters are more complicated in that the free electrons make every effort to return to the opposing battery terminal by the easiest and quickest route. Therefore, the tube analogy works if the battery is considered as a truly wonderful place, to which the passengers, if induced to leave, will want to return as quickly as possible. Furthermore, the passengers will eventually tire of all the travelling and will stay put in their marvellous retreat. When the free electrons arrive at this state, the battery is flat. In fact, a simple battery consists of two dissimilar metals suspended in a conductive liquid, the electrolyte. In the car battery, these metals are in the form of an unstable lead paste which changes chemically into an equivalent metallic substance, lead sulphate, as the battery discharges. Recharging the battery reverses this process, although a battery which has been left unused for some time will suffer the conversion of its lead sulphate into a stable, coarser sulphate which resists charging.

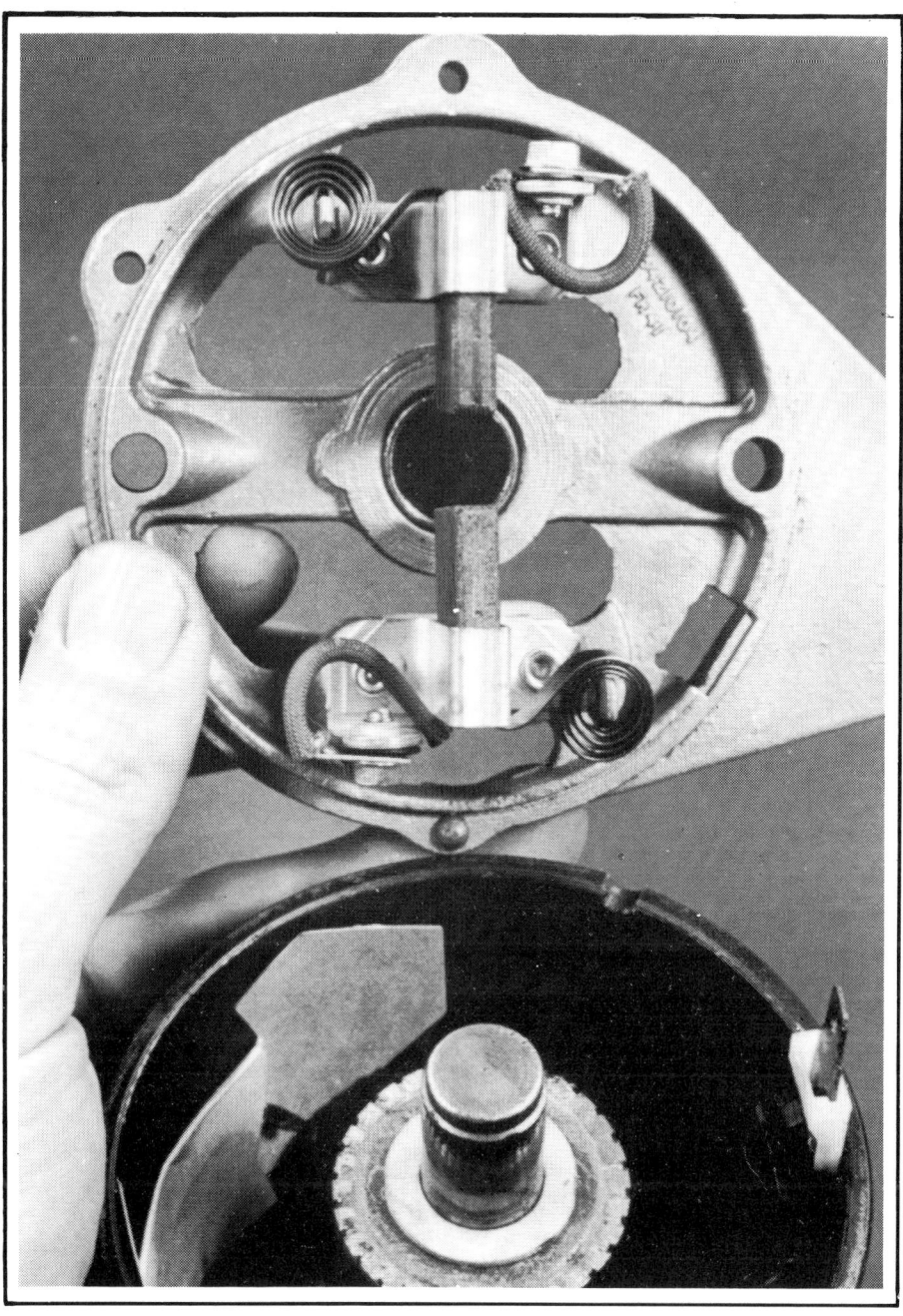

THE IDIOT'S GUIDE TO ELECTRICS — PART 2
Having introduced you to the mysteries of electrics last month, David Hill delves into your components.

The business end of a dynamo. This is a Lucas B90 exchange unit. Note the new brush holders and insulation.

The same, ready for use. The field terminal, used in polarising the unit, is the smaller of the two.

Car batteries are subject to compromise by limitations of space and weight. Their capacity must suit the demands made on them. This is expressed in terms of an Amperes per hour (Ah) rating. A 45 Ah battery would provide a 45amp current for one hour, or a 4.5amp current for 10 hours. Since the average 1500cc engine makes a demand of as much as 450amps to start it, the reason for the familiar churn, churn, silence on a winter's morning, with the battery half dead with the cold, can be guessed at. Preventing such irritating failures is easy. Dirty terminals can rob the system of significant amounts of power. Ensure that they are clean and keep them that way with a smear of Vaseline. When charging occurs, heat is created causing evaporation of the electrolyte. This concentrates the sulphuric acid solution, making topping-up with distilled water an important task. Tap water contains impurities which will reduce the battery's efficiency in time, so top up with the right stuff, to just cover the plates. In modern batteries condensation traps make for little or no topping-up, leaving an occasional recharge as the only further maintenance required. The fast, or boost chargers, employed by garages are useful things but a slower trickle charge is less traumatic for the battery. Given a healthy electrical system, only occasional recharging should be necessary, although a standing car's battery will slowly lose its charge.

Charge!!!

Electrical loads, of course, will flatten the battery in time. The logical answer to this is to balance the battery's output with a slightly greater input. This is provided by the charging system, at the cost of a small amount of engine power taken via the fanbelt. The traditional dynamo has now been more or less universally superseded by the alternator which is considerably more efficient. Both make use of a coil and a magnetic field which move in relation to each other to produce electricity. The dynamo, however, uses a static magnetic field in which the coil rotates

whereas the alternator uses a moving magnetic field within a stationary coil.

For all its superior output, the alternator has a distinct snag. The power it produces is of the wrong type. Car electrical systems work on direct current, or DC where the current flows continuously in one direction. The alternator's name comes from its property of producing alternating current, or AC, in which the current flow reverses at a given rate, in the same way as your domestic mains electricity. The problem is solved by a rectifier within the alternator. This uses four diodes, arranged in a square formation. Diodes will allow current to pass in one direction only and, in the case of the rectifier, the parallel diodes, ie. the ones on opposite sides of the square, work in pairs. The remaining pair of diodes deals with the next pulse of current. This square dance yields a continuous flow of direct current, meeting the requirements of the system. The efficiency of the alternator would lead to many a fried battery were it not for the presence of a regulator to curb the unit's enthusiasm. This regulator may be mechanical or electronic and may be mounted either inside the alternator or in a separate location. The magnetic field which induces the current is produced in the alternator's central rotor. The surrounding coils, being static, constitute the stator. These two are generally reliable in themselves but the power path which creates the magnetic field in the rotor must be up to scratch. This power is taken from the battery, so moving contacts, in the shape of the brushes and slip rings, are obviously involved. The brushes are usually held in a carrier, the unit being replaced as a whole when the brushes fall below the specified length or give trouble. This minimum length is usually to be found in your workshop manual. The slip rings are simply circular contact bands which, unless badly scored, will respond to cleaning with a solvent, such as meths. Further alternator failures tend to involve the windings and/or the diode pack. This latter can be bought separately, although it is wise to compare the price of an exchange reconditioned unit,

Gamages may be long gone, but their battery charger keeps on 'trickle charging' and is kind to the battery.

This Ford panel light dimmer is a typical variable resistance. The copper contact brings more of the current resisting wire into play. Your fuel tank sender works in precisely the same way, activated by a float.

which will also have new bearings and windings.

The alternator's predecessor, the dynamo, also produces alternating current, which may come as a surprise. However, rather than continuous slip rings, the dynamo has a segmented commutator which acts as a type of rectifier. The current is induced in the spinning central winding which, to confuse matters, is called the armature. The magnetic field is generated in the static coils which surround the armature. These are the field coils and are mounted on iron cores which exist under the resounding title of pole shoes. Brushes are also involved, bearing on the commutator to carry the dynamo's output.

Dynamos also differ from alternators in that they are connected 'against' the battery. This is not as silly as it sounds because the arrangement provides a balance between the battery and the dynamo. Thus, when the battery voltage is higher than the dynamo's output, charging stops, to restart when the battery voltage drops. This leads to a further problem. If this interchange of electricity were left unchecked, a fully charged battery would try to turn the dynamo into an electric motor. Conversely, putting the full power of a dynamo into a dead battery would cremate the latter in no time. This explains the presence of that mysterious black box, the cut-out and regulator, which prevents the aforementioned disastrous events in their respective order.

Dynamos require rather more maintenance than alternators. The commutator represents an indirect connection carrying a heavy current, leading a hard life as a result. This, and the mechanical wear involved, can cause the commutator to become scored and burnt. Sometimes, cleaning with the fine abrasive paper will suffice. Otherwise, the commutator must be skimmed on a lathe. This leaves a fine surface but the insulation between the segments must be undercut, using a ground down hacksaw blade to groove the insulation down to 1/32in beneath the surface of the copper. Brush replacement is relatively simple, although the brushes should be bedded in by wrapping the commutator with a strip of fine abrasive paper and rotating it by hand between the brushes. A deftly administered rubber band will serve to hold the brushes in against their springs during final assembly. Once again, service exchange dynamos are available at a competitive cost. A brand new dynamo must be polarised when it is fitted to create some residual magnetism in the soft iron cores which carry the various windings. Should this magnetism be incorrectly polarised, as might happen when a dynamo is transferred to a negative earth car from a positive, the dynamo will fail to work. This polarisation is carried out by touching a wire from the battery's output terminal to the dynamo's field terminal, with the dynamo fitted but not connected.

Maintenance of the cut out and regulator is limited to ensuring that its connections are good and occasionally cleaning the contacts. Inside the regulator are two or three small coils, depending on the type. Each coil, or bobbin, has contacts either above it, beside it or, in some cases, both. These should be cleaned with silicone carbide paper, with the exception of the cut-out contacts. These are made of silver rather than tungsten and should be cleaned very carefully with a fine grade of glasspaper.

Numerous tests can be applied to charging circuits to find out why the warning light is fixing you with a glassy stare and, in the case of the cut-out and regulator, adjustments can be made. These are beyond the scope of this particular feature, if only for reasons of space. Any auto electrician will carry out these procedures for a small fee, although there are books on the subject which may be consulted. At this point, it is worth noting that running an alternator with its wires removed will pass a reverse current through the diodes. The diodes will happily cope, for about 1/10th of a second, after which there will be an entertaining but expensive popping sound. Similarly, routing the unfettered output of an alternator through anything but a very large ammeter will leave you with a very burnt out instrument.

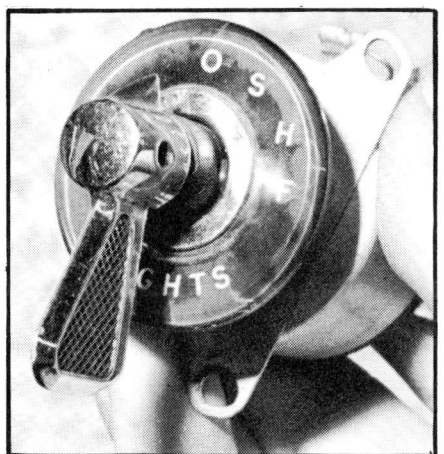

This hefty Jaguar light switch is tough enough to work without a relay. Equipment of this quality became less common with the passage of time.

This is a normally closed relay. Passing a current through the coil creates a magnetic field. The rocking arm is attracted, opening the contacts which are at the top left-hand side.

The starter forms the remaining part of the electrical system's heavy metal. Whether of the pre-engaged type or the simpler inertia variety, the starter is simply an extremely powerful electric motor. A typical starter is capable of producing more than 1bhp, sufficient to swing a reluctant crank, but it must be in prime condition to do this. So, in electrical terms, what ailments befall starters? Fortunately, 'not too many' is the response. Like dynamos, starters suffer brush and commutator wear, warranting similar remedial treatment. Some starters, however, have a face commutator. This is best considered as a being more like a disc than a drum in shape and cannot be skimmed, cleaning with meths being the only maintenance possible. Otherwise tragedies like short circuits, insulation failures and damage through over-revving are beyond DIY help, so it's back to the exchange reconditioned unit.

The thick cables used to connect starters also explain the necessity for the solenoid. Electric motors work by harnessing the power of electrically created magnetic fields, in particular magnetic repulsion. Solenoids use magnetic attraction to move an iron core within a coil. The question is, for what purpose? The answer lies in the thick cables. Without a solenoid you would be using a Dr. Frankenstein-style blade switch to work the starter because the more familiar type of switch would disappear in a blaze of annihilated contacts under such heavy currents. When you turn your key to 'start' you are energising the coil in the solenoid. The coil attracts the iron core, closing a pair of heavy contacts which pass current to the starter. The solenoid may be a piggy back unit on the starter, or it may be a remote unit; the principle is the same. By the same token, the electric fuel pump is a repeating solenoid, attached to a diaphragm.

We have ignition

The property in which the movement of a magnetic field will induce an electrical current in an adjacent wire crops up over and over again. We have Michael Faraday to thank for this discovery which is, in fact, the law of electromagnetic induction. The same law applies to the ignition coil, although the movement involved is not a visible one. Passing a current through a coil of wire generates a magnetic field. As the current flows, this field spreads outwards. This spreading constitutes the necessary movement, inducing a current within the very wire which has caused the magnetic field. Switching off the original current which created the magnetic field (with me so far?) causes that field to collapse instantly. Here, the word 'collapse' can be taken literally. The magnetic field subsides like a pricked balloon. Since another relative movement between magnetism and a wire is taking place, another current is induced. The difference is that the short time involved makes this a very powerful current, say around 300V from a 12V input. Placing another coil, but one with many more turns of thinner wire, within the first coil has an interesting effect. Although the inner coil is not physically connected to the outer it will pick up the effects of the collapsing magnetic field and magnify it to as much as 20,000V. This voltage is capable of jumping gaps, as any sparking plug illustrates.

In a car the contact breaker, as its name suggests, opens the circuit which collapses the magnetic field in the coil. However, the aforementioned voltage can jump the contact breaker gap. Apart from burning the contact breaker points away, this robs the spark plugs of power, which is bad news. This is prevented by the condenser, which can store an electric charge. During the instant when

The Lucas three bobbin dynamo control box is often encountered. The bobbins are, reading from the left, the cut-out, the current regulator and the voltage regulator. The disc-shaped objects at the rear are the cams which alter the contact tension and therefore the rate at which the contacts operate. Ill-informed adjustment is not recommended.

the points begin to open, the potential leaping spark enters the condenser in the form of an electric charge. The condenser, thus filled, develops a voltage equivalent to that of the spark, preventing any further charge from flowing. The condenser thus acts as a variable buffer, taming the errant spark.

Cold starting efficiency is the watchword of the ballasted ignition coil often encountered in later classics. In cold conditions, a big, healthy spark at the plugs is useful. The ballasted coil provides this in a simple but effective manner. In a normal system, the coil runs on 12V. A ballasted coil is designed to run on 6 or 7V, with a resistor turning the remaining voltage into heat. During starting, the coil is fed with 12V from the ignition switch's 'start' position. Those extra volts cause a bigger spark, the better to fire up the cold engine.

Telling people of your calibre how to maintain your ignition system would insult your intelligence. However, the purpose of the exercise is worth knowing. Contact breaker gaps close up in service, owing to the wear occasioned on the heel by the rotating cam. This shortfall in the gap may seem beneficial in allowing the coil a fraction more time to induce a better spark. In practice, the efficiency of the contact breaker's switching is adversely affected, which explains why the resetting of a badly closed points gap can dramatically restore the engine's power.

Electronic ignition may be encountered in the later classics. This system offers distinct advantages over conventional ignition. Despite the condenser, contact breaker points will suffer from erosion through arcing. The simplest type of electronic ignition uses the contact breaker as a signalling device. As such, the points are required to handle a mere 2% of the current involved in conventional ignitions, hence the vastly improved points life associated with transis-

tor-assisted ignition systems. Beyond this point, electronic ignition becomes complex in the extreme. To describe inductive discharge systems, capacitor discharge systems and the Hall effect (which is presumably better than the living room effect) would require a full magazine's worth of space. More to the point, the aspects of maintenance are limited. Black boxes and alchemy are involved and replacing faulty components *in toto* is easy enough to do. Having said that, contactless distributors still employ conventional bobweights and vacuum advance mechanisms, which can be maintained as per those in a mechanical distributor. What is worth knowing is that capacitor discharge (CDI) ignition systems are capable of immense spark voltages. A conventional system's high tension shock will make you feel uncomfortable; a CDI shock could kill you.

Inside story

Much of the car's electrical system lurks unseen among the grubby mechanicals. The components inside the car put you in control and the congregation of connections in the dash area is a matter of necessity. Switches are a case in point. Although they are but a simple mechanism, they are more important than meets the eye. If that seems a little high-flown, imagine yourself being plunged into absolute blackness by a switch failure while approaching a country road bend at speed on a rainy night.

Fortunately, these devices live in a cosy location, with only mechanical wear and security of connection to compromise their operation. For those functions involving a heavy current, relays are used. A relay, like a solenoid, is a remotely operated switch, using a magnetic field to close a pair of contacts. They are also easily maintained, usually being opened only when dirty contacts cause problems.

Instruments lead as easy a life as switches, which is fortunate because they are relatively delicate. Three main types are used, their occurrence depending on the car's age. Originally, moving iron instruments were in virtually universal use. In these, the indicating needle is attached to a small, freely pivoted iron block. A pair of coils, mounted within range of either end of the block, provides a magnetic pull which varies in relation to the information received from the transmitter. This transmitter provides a variable resistance to current, balancing the relative pull of the two coils. The next type of instrument works with the aid of a bimetallic strip. For those who spent their physics lessons asleep or watching the football outside, a bimetallic strip is made of two metals joined firmly together. The metals have different rates of expansion, so that the strip bends when it is heated. Bimetallic strip instruments have a U-shaped strip, with a tiny coil of wire wound around one leg. As current passes though the coil the strip is heated. The other leg of the U carries the indicating needle so, as the strip responds to being heated by bending, the needle moves. Both moving iron and bimetallic strip instruments must have a constant incoming voltage if the readings are to be accurate. This constant voltage is provided by an instrument voltage stabiliser, in which a bimetallic strip controls a set of contacts which feed its own heating coil. The contacts open and close in a continuous cycle, yielding a steady 10V or so. A bimetallic strip is used in a similar way to make your indicators flash.

The final type of instrument is a recent development and needs no voltage stabiliser. This is the air-cored instrument, in which the pointer is attached to a small magnet. Three coils of wire surround the magnet, each coil providing a magnetic field in a different plane. Keeping the voltage in one of the coils constant, while varying the voltage in the other two, causes a proportional variation in

A close-up of the cut-out contact shows that this, too, is adjustable. This is the soft, silver contact which requires care in the cleaning.

A Lucas alternator stripped for action. The finned item with three spade terminals is the rectifier pack.

within a glass envelope. Without this, the filament would vaporise in short order.

In every case, a circuit is involved, whether it is a complete circuit of wire, or a circuit returning to earth via the cars' conducting shell. The consequence of this is that most of the maintenance required is concerned with keeping those circuits complete, whether through wires, connectors, carbon brushes or whatever.

A time exposure shows a bimetallic strip instrument in action. The U-shaped strip and its heating coil are clearly visible. The needle took 8 seconds to traverse the arc, illustrating the lack of a need for damping. Note the delicate nature of the construction.

the attraction of the needles' magnet, with a consequent accurate movement of the needle.

• In respect of the oil pressure and water temperature gauges, two types of transmitter are used. The temperature transmitter contains a thermistor, or semiconductor resistor, whose resistance to current falls as the temperature rises. The oil pressure sender uses yet another bimetallic strip, whose heater coil contacts are controlled by a flexible diaphragm. The rising oil pressure makes the diaphragm close the contacts, sending a current through the heating coil. This makes the bimetallic strip bend to open the contacts once more. The process continues as a cycle until a balance is reached, the result being indicated by the gauge. All these resisting senders cause a high reading to occur when the resistance is at its lowest, which is particularly handy. Removing the sender lead and attaching it to earth will cause the relevant instrument to give its highest reading, which is a good way of deciding whether the instrument or sender is causing a fault.

Tachometers of the electronic variety are a special case, having an internal circuit board sensitive to polarity. This is reason enough to mark the connections of your tachometer; connecting it backwards will cause its instant demise. The tachometer measures and displays the frequency of the pulses caused in the ignition supply wire by the coil's being triggered. This explains the apparently useless loop of wire which can be found on the back of some tachometers. This is the induction loop, which is actually part of the feed wire from the ignition switch to the coil.

The simple answer

The foregoing illustrates the fact that the majority of electrical components use the same basic principles of magnetic induction and repulsion, and generated heat. The same applies to such items as the horn, the heater motor and the rear window demister. The wiper motor is also part of the scheme, with a simple internal override switch providing a current to park the blades. The car's lights use generated heat to a high degree, the thin lamp filaments being made white hot by the passing current, hence the need for a vacuum

Since this is the idiot's guide to electrics, the intense complications of current consumption, back electromotive force and the behaviour of magnetic flux have been avoided. Being informed, rather than confused, yet sufficiently well-armed to do battle with your electrical system, has the fortunate effect of removing some of the mystery. Fear of the unknown is a powerful enemy.

The third and final part of this meisterwerk concerns the making of a wiring loom from scratch. This is less daunting than it sounds, as will be revealed next month. In the meantime, solder on! □

My thanks to Lucas agents, York Autoelectrics, for the loan of new electrical equipment for photography.

AUTOELECTRICS

Those of you who have read, marked, learned and inwardly digested parts 1 and 2 of this magnificent production will nearly be experts on vehicle electrics. I say nearly because this last instalment must be absorbed before students become eligible for the order of the golden spark plug.

The purpose of this episode is to instruct you in the art of making a wiring loom, which is a perfectly possible DIY job. What's more, the modest admission that you did your own wiring not only earns open admiration in clued-up company, it has also been known to cause the odd commission to make another loom. This happens because most people, as I have said before, are afraid of wiring. But that's only fear of the unknown, into which we now embark.

Clamp type battery terminals are much more reliable than the cover type. These are better still, being properly sized for positive and negative terminals.

Where to start?

Where indeed? Well, the need for a loom will have been dictated either by the absence of any loom whatsoever, or by the generally tatty and/or non-functional state of the original wiring. Beyond revealing the original position of the wires, the old loom is to be discarded but, in the event of the inability to track down a rare connector, it is wise to keep the old loom for now. Better that than a complete halt for the sake of one small part. In the original vehicle there will be clips, or at the very least their spoor, to indicate where the old loom went and, for measuring purposes, all of the electrical units should be in place. Thus equipped, you can use a flexible tape measure to discover the length of each run of wire. This causes a snag, in that the various colours of wire must be known before the length of each required can be calculated. A British Standard colour code exists, to which the vast majority of British cars conform.

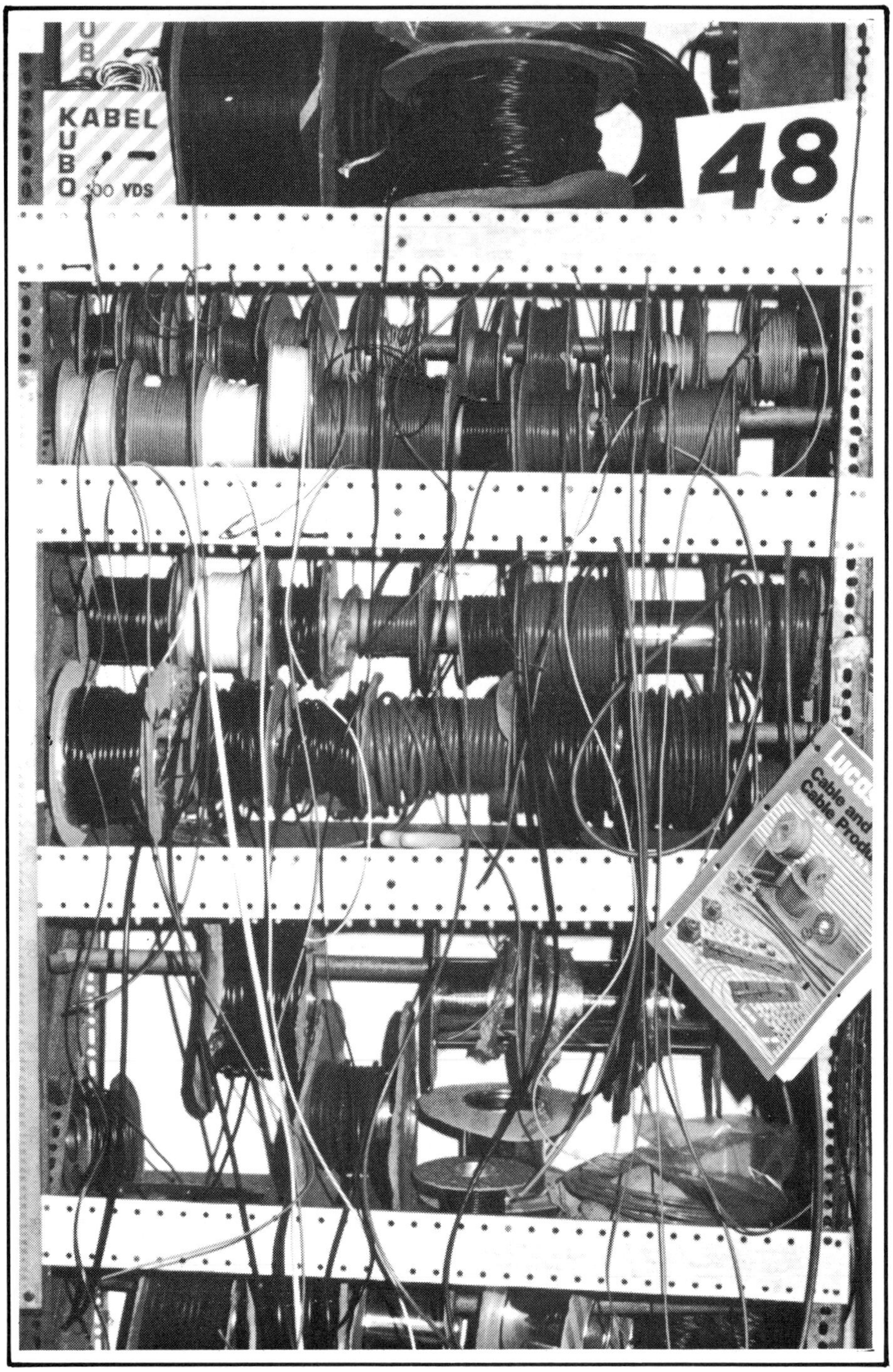

THE IDIOT'S GUIDE TO ELECTRICS — PART 3
In this third and final instalment, David Hill makes it all happen.

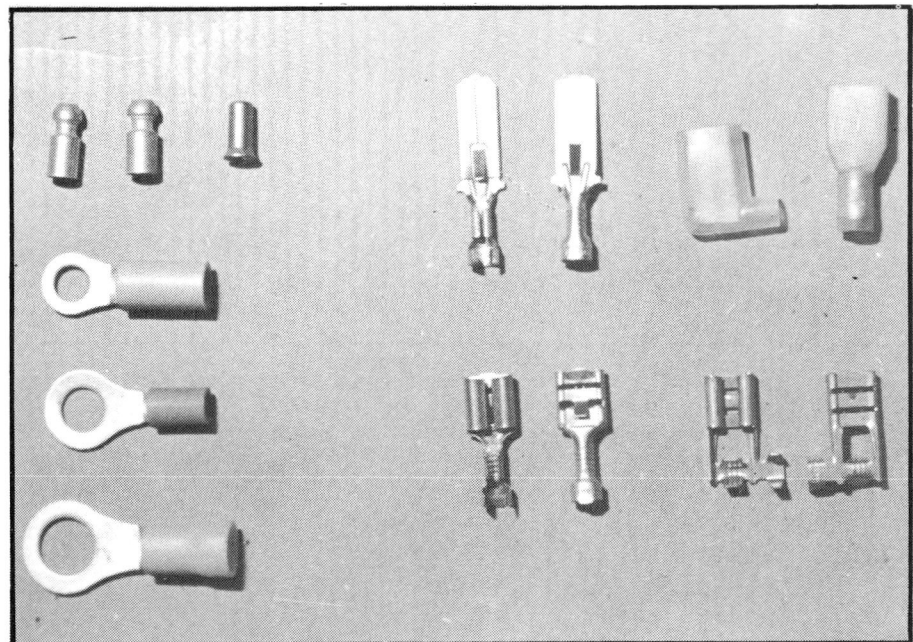

These are the type of terminals you need. The ring terminals on the left are the crimped type. Better to use the solder type for reliability.

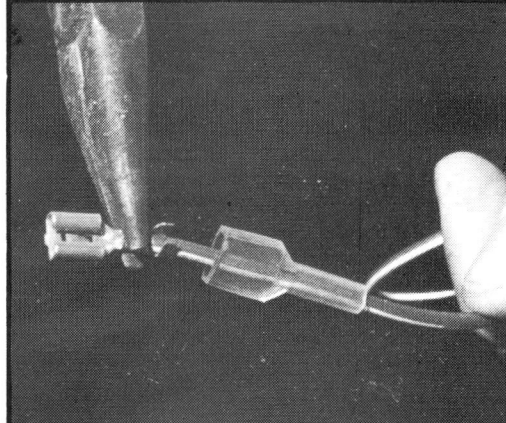

Fitting a spade terminal involves crimping the smaller of the two sets of claws on to the stripped wire. Note: the sleeve has been fitted before the terminal.

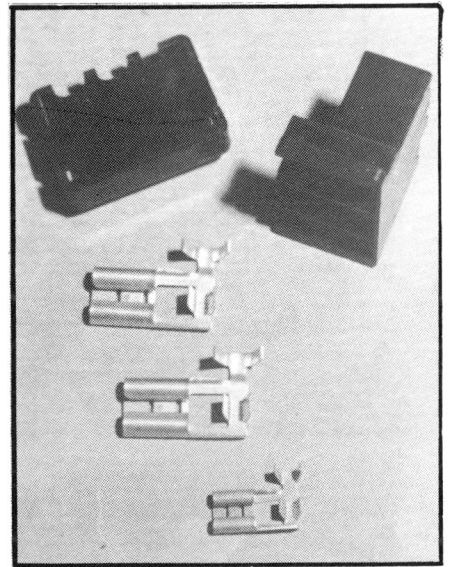

Lucas Autocentres sell these alternator terminal blocks as a kit. Using this is far easier than making do with an old block.

However, quoting this in its entirety would not only be confusing, it would also be irrelevant. The reason for this lies in the code being fairly recent and therefore covering accessories which may or may not be present in your car.

The easy method is to make use of the vehicle wiring diagram, examples of which are to be found in the various types of manual. Armed with this, you can take each wire in turn and, having learned its starting point and destination, you can physically measure that run on the car. For example, the headlight main beam circuit involves a run from the main lighting switch, via the dipswitch to the front of the car, splitting into two to feed each light unit. This is done using blue wire with a white trace, usually termed blue/white, the second colour being the trace colour. Conveniently, the dip beam circuit is the same thing in blue/red.

Further generalisations are possible. Those circuits which are alive only when the ignition is on are hooked up with plain green wire as far as the switch. Then a green wire with a trace runs from the switch to the unit. For instance, the heater blower is fed from the 'ignition live' section of the fusebox to the switch in green, then a green/yellow wire goes from the switch to the blower motor. Unfortunately, there are foreign cars, notably some French and Italian models, whose looms use about three colours. This causes lots of fun when something fails and my advice when rewiring is to use BS colours. Alternatively, you could spend the wait for the AA man in congratulating yourself on having an original loom.

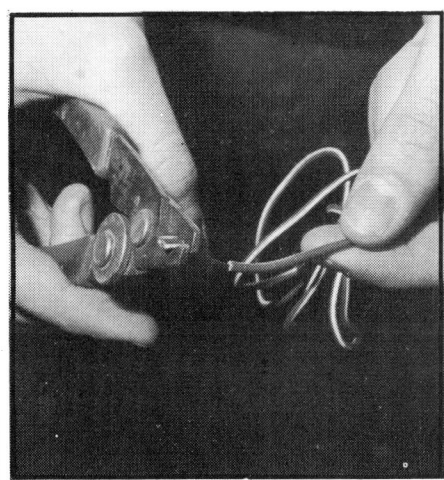

A good pair of wire strippers won't cost the earth and help enormously. These can be set to suit the wire size.

Weight problem

Using the wrong weight of wire is a trap into which the budding autoelectrician tends to fall with a resounding clang. Too thin and 'hot wiring' a car takes on a new meaning altogether; too thick and you will have an ungainly, hard to handle loom. Cable weights are expressed in terms of numbers. The first number, before the stroke, refers to the number of strands in the wire. The second number refers to the thickness of each strand, in imperial or metric terms. Thus 28/0.012 wire has 28 strands of 12thou wire. The metric equivalent is 28/03. As a guide, the following wire sizes will suit the following jobs:-

14/0.012in (14/0.3mm). General wiring for such as indicators, sidelights and radios.
28/0.012in (28/0.3mm). Heavier wiring for headlamps, heated rear windows etc.
44/0.012in (44/0.3mm). Dynamo output.
65/0.012in (65/0.3mm). Small alternator output.
84/0.012in (84/0.3mm). Large alternator output.

With this information to hand you can make a full list of each wire required, with its length, weight and purpose. In measuring, absolute accuracy is too much to hope for, so add at least 6in to your measurements to allow for connection and clipping.

The next step is to track down a sales outlet with a full range of wire colours. This is not as easy as might be thought. The major outlets like Halfords sell single colour wire on reels but, for trace colours, large auto electrical suppliers like Lucas Autocentres are the places to look for. Under no circumstances use anything other than correct car wire; some outlets sell steel wire. It will fracture and let you down.

Making connections

While buying your wire it is a good idea to pick up the necessary connectors. A vexed question arises concerning the type of connector to use. Many advocate the use of crimp-type connectors, identifiable by their bright insulating sleeves. Personally, I don't trust them. They may be quicker to fit but,

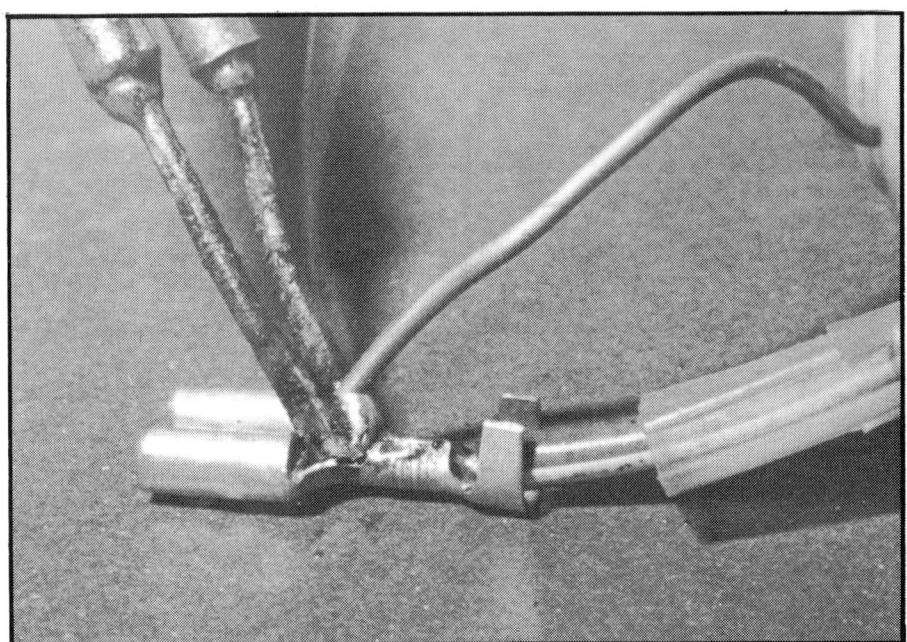

Solder is used to fill the channel the spade and make a tough, electrically sound joint. Take care not to use too much or the terminal blade won't go in.

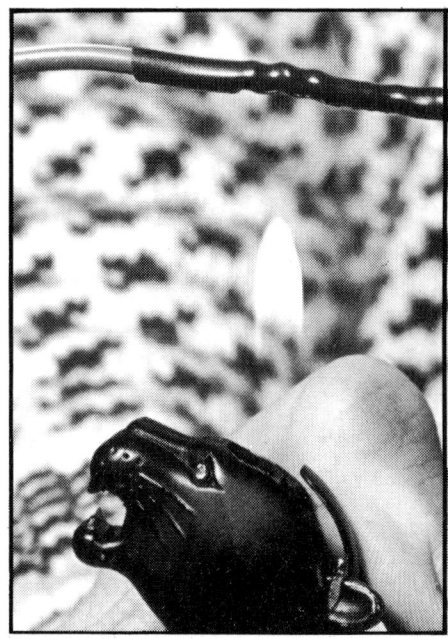

Heatshrink tubing will insulate the joint properly even if you don't have a panther-shaped lighter. Radio shops sell heatshrink in a variety of sizes.

for my money, they are too likely to lose their grip unless fitted very tightly. In this latter case it is all too easy for the copper strands inside to fracture, with the same result. Since these crimped connectors are also non-original, I suggest the use of more traditional, solder-type terminals. These can be divided into three basic types: Lucars (or spades), ring terminals and bullets (or snap connectors). The first two come in various shapes and sizes, and there is a variety of sleeves for the latter. A quick head count on the car will give you an idea as to the amount of each required but you can often buy a quantity of such items at a discount. Don't forget to buy the insulating sleeves for the spade terminals as well. You will also need some solder, preferably of the multicore type used in electronics, and some flux. A standard paste flux such as Fluxite will do the trick. Never use acid flux, which is for use in soldered fabrications like radiators. Acid flux will allow a joint to be made but will continue to work on the copper inside the wire. There is no way to prevent this from happening because the flux is pushed along the core of the wire by the heat. Eventually, the acid eats through the wire and the car ceases to work.

Next on the list is a means of protecting the wires and their joints. In certain areas joints have to be made in mid loom, such as where a wire splits to go two ways. Heatshrink tubing, as used in electronics, is useful for this, as well as for sleeving the shanks of ring terminals. Radio shops sell it. As with terminals, a difference of opinion relates to the taping material used for looms. Some maintain that plain plastic tape should be used, with self-adhesive tape at the ends only. This plain tape is available at your sales outlet. For my part, I have always used sticky tape as a wrapping and, with more than a dozen looms in use, have never heard of any problems at all.

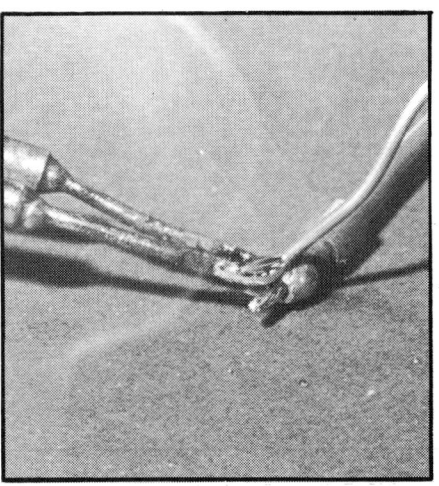

The secret of attaching a bullet is to get the terminal and the wire equally hot. The joint happens inside the bullet. Note that the soldering is being carried out on non-conducting cardboard. Scrap wood serves equally well.

Twisting wires together is perfectly acceptable, provided that they are soldered too. Note the solid sheath of solder to the left of the iron ⌐ this joint won't come apart.

Unless someone can fill me in on the reason for using non-stick tape, I will continue to use the other stuff.

The penultimate entry on the shopping list is a soldering iron, unless you already have one. This type of work demands a fairly small iron, electrically heated devices being far more convenient. A size of around 30W will do very nicely. Personally, I use an instant heat soldering gun made by Bosch. This heats up in seconds, is powerful and even has a little searchlight on the front. A good wire stripper is also handy to have around. Lastly, there are the bits and bobs to hold the wires in place, and the special terminals. Unless your car's cable clips are re-usable, you will have to buy new. Various sorts are available, even the original types. Buy enough of the right sort is the hot tip. I have little respect for the stick-on type but recommend those nylon ratchet clips known as Tywraps. In particularly hostile areas a similar-sounding product can be used. This is called Spiwrap and is a tough, flat nylon tape wound into a spiral. For areas like behind the bumpers and where the loom passes over chassis rails and the like, it is virtually chafe-proof. Naturally, people of your calibre wouldn't put wires through a panel without the correct open grommet — would you?

The special terminals mentioned include such as the special blocks used in headlamps and alternators. The old ones could be used but it's very awkward, what with years of corrosion to be removed before soldering. Other specials are the right-angled type of spade connector, male spade connectors, non-solder bullets for lamps and all conducting, plus triple sleeves for bullet connectors. These were included in the original system for a reason so put them back where they should

AUTOELECTRICS

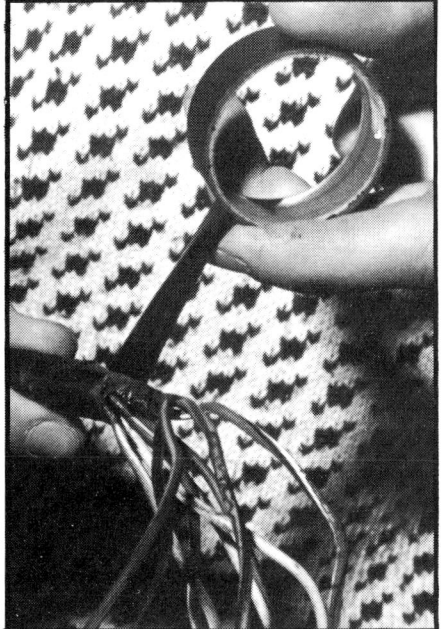

Taping the loom is time-consuming, but must be done properly. Keep the tape taut at all times.

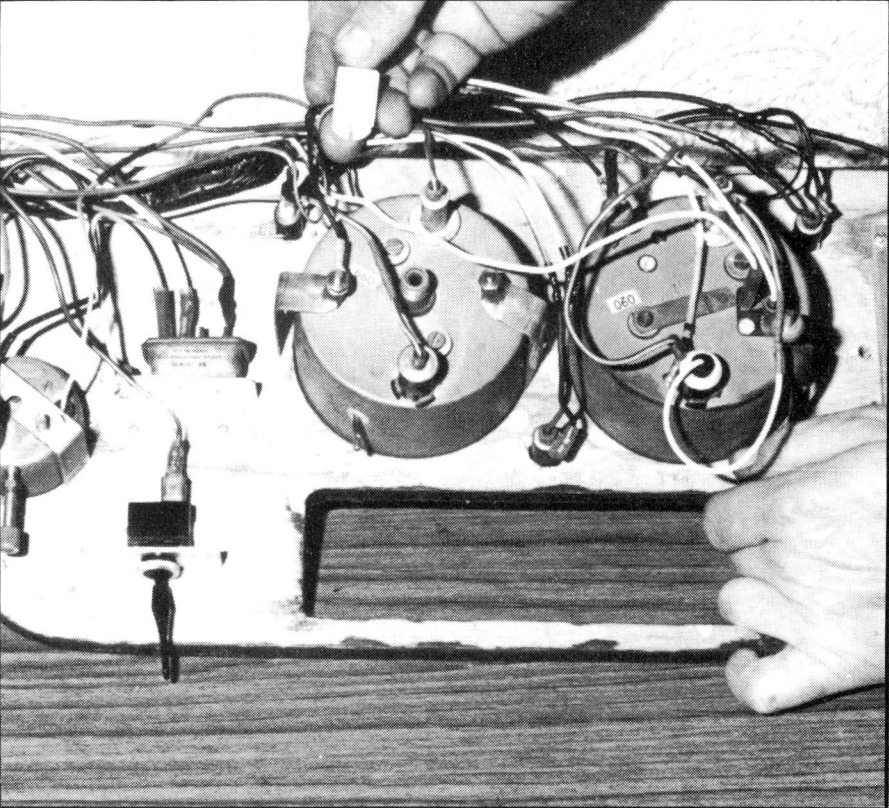

....And here's one I did earlier. This is a kit car loom but the principle is exactly the same. The wires have been laid and connected, ready for final taping. Note the voltage stabiliser for the instruments, just above the switch.

be. In certain cases life can be made easier by secreting a multiplug in the loom. It may be non-original but, if hidden away, it can make removing a sub-loom, such as for painting the engine bay, a lot simpler.

Rigging for victory

Before plunging into the fray, a word to the wise about soldering wouldn't come amiss. Cleanliness is the secret; dirty materials won't solder. Given new wire and new terminals, this shouldn't be a problem provided the tip of the iron is given a clean from time to time. Any joint is prepared in the same way. Physically assemble the joint, adding a touch of flux. Dip the hot soldering iron into the flux tin and put a little solder in the tip. This will spread and 'tin' the iron; if it doesn't, give it a wipe and try again. Then dip the iron in the flux once more and touch it to the work. When the flux on the work is bubbling nicely, feed in the solder until you have completely filled the joint. Let it cool, give it a tug and, if happy, wipe over the joint and it's done. If the melting solder doesn't run like mercury there is insufficient heat and the joint will fail.

Now comes the business of wiring the car. The solenoid often constitutes the beginning of the loom so it is a good place to start. From there, work to the fusebox, establishing the ignition live and permanent live sections. Then work through the system, circuit by circuit. With the clips in place you can lay in each wire, leaving a long tail by the starting and finishing point. Before long you will have a respectable bundle to deal with, which is where confusion can set in. The answer is to tape each set of wires together. You can do this as many times as you like, since the wrapping will hide the taped points. The procedure is to lay all the wire before a single termi-

nal is fitted. This may seem an illogical way to go about things. However, there is a reason. The seemingly obvious method of connecting one end, then the other, then wrapping the loom, causes problems because the overall length of the loom reduces when the wrapping goes on. My first loom suffered this shrinkage and had to be remade. If yours does the same, don't blame me. Further confusion, at points like the dash, can be avoided by temporarily labelling the wires.

Power crazy

With each and every wire in place, you can make up the initial taped bundles. Wrap the wires tightly with sticky insulating tape every three inches. Avoid stretching wires or making them strain in the clips. If it's wrong, cut the tape and do it right – there won't be another chance. These actions will have the wires looking something like a loom. When all the wires are neatly placed, you can start fitting terminals. Place each wire end by its destination, mark the spot, add two inches and cut. Then strip the wire end, fit and solder the terminal and fit it on to its location. Do the same for each connection until the whole of the loom is connected up.

The good bit comes now – testing your work. If your organisation is worthy of the title, the system should work perfectly. Having said that, I wouldn't connect the battery tightly at the first go, nor should you. If something fails to work, trace the cable through until you find the fault. If it says

'Ffrrz' there is a short circuit and the position of the exterminated fuse will narrow its location down for you. The common silly tricks are connecting the ignition live to the points instead of to the coil, forgetting to attach a terminal or two and expecting the solenoid to work without an earth. Oh, and yes, ballast resistors do get very hot in service, especially the ceramic type. Wiper motors also cause problems, what with two speeds and electric parking devices. Given that you know which is the earth terminal, you can swap the others around at will until it works.

It's a wrap

The testing portion of loom making is the reason for leaving the wires unwrapped until now. Imagine discovering that you should have made a two- way split after the wrapping is done. The de-bugging may take some time, but don't contemplate the wrapping until it's perfect. Then, you can finish the job. The loom doesn't have to come out altogether to be wrapped but you can remove it if you prefer. In that case, temporary labels on the wires will help.

The business of wrapping is more complex than meets the eye. The aim is to integrate the wires into a tough yet flexible whole. Making the free ends secure is therefore a good thing to avoid flapping bits of tape. If the loom is considered as a tree branch, you start with the twigs. Leave three inches of wire at the terminal and before you start wrapping or there will have been no point in

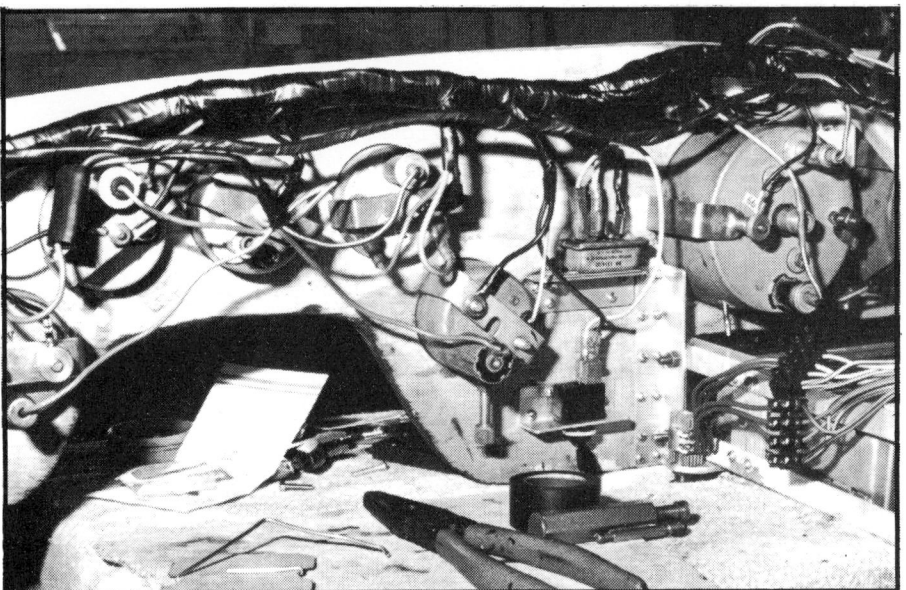

The finished job, prior to installation in the car. With such a close arrangement, clips were unnecessary. Note how every live terminal is sleeved. If one should become loose and turn it couldn't short circuit and cause a fire.

using coloured wire. Working left- or right-handed as preferred, wrap the tape around four times. Then, keeping the tape taut, set off down the wire, overlapping by half the width of the tape. When the main loom is reached, imagine you are bandaging a thumb, going round the main loom, back round the incoming wire then round the main loom again. Lastly, cut the tape at an angle, which will make it less likely to come unstuck as time goes by.

This will leave you with a bare main loom with all its incoming wires taped. One of those wires will constitute the end of the loom which is the starting point for the main taping. Wrap the loom from the end to the middle, repeating the exercise from the other end. This leaves the vulnerable join in the middle, at an undisturbed point beneath the dash. A correctly taped loom will have uncovered wire sheaths only at their termination points. If there are gaps, wrap them, especially in areas outside the passenger compartment. Finally, refit the loom, test it once more and that's it.

Job satisfaction apart, making a loom in this way will not only educate you in the ways of your car's wiring but it will also save you a lot of cash. Also, with properly soldered terminals, there is no reason why electrical failures should happen, which is reason enough for taking the trouble.

The perversity of electricity in relation to cars undeniably causes trouble but it is hoped that this trilogy has stripped away a lot of the mystery. The electrical system is a mechanism like any other; don't let it scare you. ☐

Last of all, I would like to thank Andy and Chris at Lucas Autoelectrics, for their invaluable assistance in the preparation of these features.

96

A SPARKLING COLLECTION

A car-related hobby with a difference. By Don Cruickshank.

Have you ever thought of collecting spark plugs? Probably not but, before you dismiss the idea out of hand, just let me demonstrate the scope of the subject. If it was decided to collect only one of each make, past and present, the collection would run to many thousand plugs. When you consider each manufacturer will produce a range of at least ten and nowadays probably nearer 100 different plugs you will begin to see the collection could get quite large. Perhaps at this stage we shouldn't take into account that manufacturers change their range every year or so or we will be put off by the thought of such large numbers.

I think we have established that there exists enough variety of plugs to form a collection but what would be the point of it? Modern spark plugs are designed with the aid of computers that can analyse all the engine functions but, in the early days of motoring, each designer had his own thoughts of how the perfect plug should look. What one

Removing a spark plug for cleaning or replacement. Most of us simply throw away the old ones but collecting them can be a fascinating pastime.

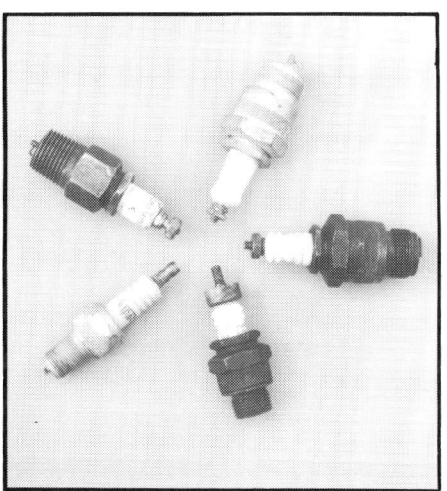

Nowadays (apart from a few examples where a smaller head is required, the FD/FE Victor is one example that springs to mind) most spark plugs are a common size and shape. In the past this wasn't so though, as this selection demonstrates.

designer was convinced was right his fellow designers would be equally convinced was wrong. This makes the old plugs a fascinating subject for study as each can be so different.

We have become used to seeing plugs with 4mm threads, which is the size most modern plugs use. Early plugs favoured larger threads and the most popular was the 18mm diameter. In the early days the 18mm thread was known as the continental thread to distinguish it from American threads. The American threads were a ⅞in diameter parallel thread and a ½in measurement refers to the nominal bore of the thread the pipe was intended for, rather than the outside diameter of the thread. This thread is cut on a taper so it doesn't need a gasket to form a seal. Now this may seem a bit haphazard because the harder the plug is tightened the further it goes into the head but then the position of the spark in the engine was not so critical as it is in a modern engine. Early engines also tended to have lower compression ratios and thus more clearance between spark plug and moving parts.

A very high proportion of early plugs could be taken apart for cleaning purposes. These

This Champion X dates from around 1919. Made in America (Champion did not start manufacturing in Britain until later) to fit early Ford cars such as the Model T, it has a taper thread known either as a Briggs thread or a ½in taper pipe thread.

plugs were known as demountable plugs. The usual form was for a gland nut, about the size of a modern lug hexagon, to clamp the insulator between two copper asbestos washers into the main plug body. The hexagon size of the main body would have been at least 1in across flats. The insulators could be purchased separately to repair broken plugs. In the early days of motoring mixture strength could be far from ideal and many engines used large quantities of oil from new so the incentive was there to make plugs easy to clean.

The fixed plug, or one-piece plug as we know it today, did not become popular until about the mid-1920s. This is not to say one-piece plugs did not exist considerably before this because they did but the market was not right for them. The one-piece plugs faced two main problems and only one of these was within the control of the spark plug industry. The engines the plugs had to fire were prone

Like several early plugs, it is demountable or detachable, ie. it comes apart for cleaning and servicing; plugs were not then regarded as throwaway items.

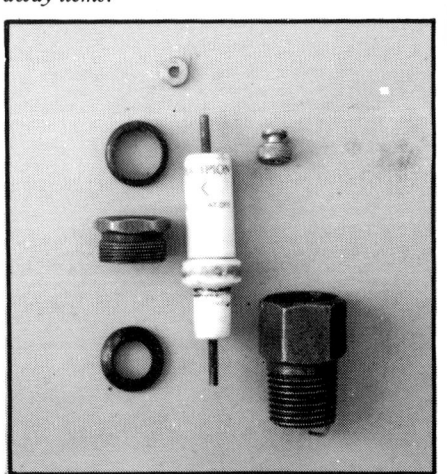

to oiling and none of the patented devices the plug manufacturers came up with could cope with this menace. The second was the method of sealing the insulator in the plug body. For many years the plug manufacturers were unable to get over this problem until insulators were made with a similar co-efficient of temperature expansion to the steel body they were clamped into. Early one-piece plugs had a bad name for blowing out or, to be more precise, blowing by the side of the insulator. Manufacturers of the then more traditional demountable plug often made a great play of this in their adverts. If a demountable plug leaked an extra nip up on the gland nut would soon cure the problem and, if the worst came to the worst, the copper washers could be annealed and the whole thing re-assembled.

We take electric welded joints for granted now but this level of technology was not available to the early spark plug designers. If you look at any modern plug you will see that the earth electrode is welded to the plug body. Now just for a moment pretend you are designing a spark plug but there is a restriction placed on the design. The earth electrode may not be welded in place. This is exactly the problem that faced the early spark plug industry. Many solutions were found, some more successful than others. Some manufacturers drilled an interference hole in the base of the plug and pushed a round wire electrode in it. This was quite good provided the plug did not get too hot. If the plug overheated, the wire was inclined to fall out and, I suppose, in the majority of low compression engines was passed into the exhaust without causing much trouble.

A slightly better idea on the same theme used an interference hole drilled radially through the threaded part of the base at right-angles to the plug axis. An electrode pushed into this hole could not fall out into the engine. If it came loose the worst that could happen was to short the centre electrode to earth. The best solution that did not involve welding was to turn a recess in the base of the plug and place a stamped-out washer, usually with three electrodes on it, into the recess. The lip of the recess was then rolled over to hold the assembly in place. This method was so successful that it was still retained long after welding technology had developed the butt-welded electrode.

For some reason, manufacturers of early plugs tended to make plugs for buses and trucks longer than those for use in private cars or motor cycles. Perhaps this was in some way dictated by the large and clumsy open-ended spanners in use at that time. Just compare an old spanner and a new one of the same end size and you'll see what I mean. The majority of early commercial engines tended to be sidevalve, with the plugs on top of a flat head, so in that respect they were very accessible. Perhaps the longer bodies were less likely to be crushed when clamped in a vice to be taken apart for cleaning. Every time I see an early illustration of a plug clamped in a vice, it makes me shudder. So many plugs have been squeezed out of round by clamping them too tight in a vice. If you have to take an old plug apart, a safe method is to use two ring spanners. Clamp one of the spanners in a vice if you like, then you will be able to apply a greater torque to the plug

A SPARKLING COLLECTION

Another Champion, this time an R1 racing plug from the 1930s with an 118mm thread; combustion chamber space wasn't so short in those days! The side electrode is set into the plug body. Notice also the brass ring on the top; this is to aid sealing and prevent blowby up the side of the centre electrode.

without distorting it. Anyway, who wants vice marks all over a plug collection?

One of the most important points to a collector must be the age of a plug but, unfortunately, it is not always that simple to judge how old a particular plug is. There are many people who will tell you that one with a mica insulator must be older than one with a ceramic type but this is not always the case. Some manufacturers started off with mica plugs and graduatd to ceramic. Others started with ceramic and then introduced mica insulators. Some manufacturers have

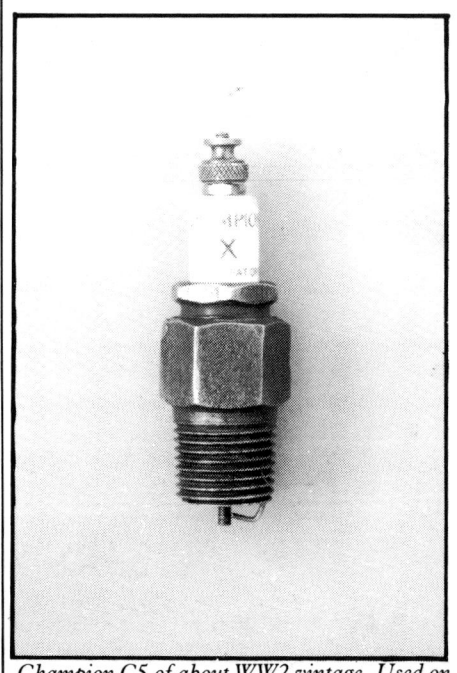

Champion C5 of about WW2 vintage. Used on many different American cars of the era.

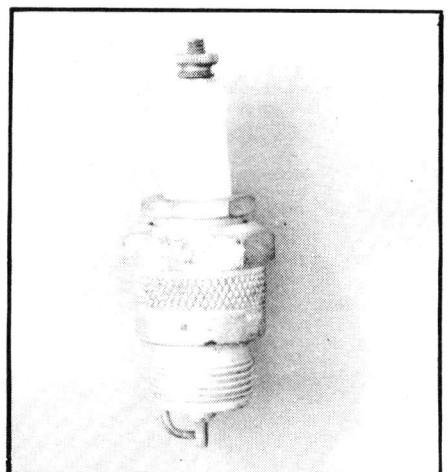

Pacy 25. Built for use with Fordson tractors by Pacy, a small company named after its proprietor.

never produced a plug with mica insulators. Leaded petrol spelt the death of the mica-insulated plug.

The lead causes a build-up of material on the insulator which reduces its electrical resistance. A friend of mine who runs a vintage car is going to turn this build-up off the insulators of an original set of mica plugs and try them with the new unleaded petrol. On the face of it there is no reason why this should not work as the original mica insulators were turned on a lathe. The general procedure was to wrap mica tape round the central electrode then force mica washers together over this tape and clamp them with a threaded brass cap. The whole assembly was then turned to size and fitted into the plug body.

A complete guide to plug dating would probably take years to produce but, with a little knowledge, quite a good estimation can be made. The clues lie in such things as the shape of the insulator and the colour or style of the writing on it. The body finish and the size of the hexagon provide the next visual clue. The type and method of attachment of the earth electrode and the size of the top terminal thread provide slightly more subtle clues. The size of the main thread depends on the model of the plug, not its age, so it does not follow that a plug with a ⅞in thread is any older than one with a 14mm thread. Original boxes provide a good means of dating plugs. Some firms changed addresses so the address on the box will give an approximate idea of age. Some pre-WWII war plugs were packed in metal boxes and the way the box was stamped out and the positions of seams all help the expert to arrive at a date. Wartime boxes tell you to save cardboard towards the war effort which narrows down the date considerably. The colours and style of printing on boxes can be compared with old adverts to provide a guide to age.

Let us assume enough interest has been aroused to start a collection. Where do you find old plugs? The most obvious place is an autojumble, where all sorts of unlikely things can turn up. Car boot sales sometimes can be rewarding if you are prepared to sort through the bottom corners of junk boxes. By the way, be careful how you clean old plugs. By all means use a wire brush to remove rust from a plug body but don't use a wire brush on the insulator. When I once did so, all the

Wipac P40L. This 14mm plug looks like anything that could be found in a high street accessory shop but in fact the insulator pattern shows it to date from the 1950s. Wipac were formed when the Pacy company mentioned above merged with Wico, thus forming Wipac although for many years they were known as Wico-Pacy.

writing came off.

I'm loathe to mention price when it comes to old plugs but let me put it this way. With a bit of luck and £10 to spend at an autojumble an embryo collection could be formed. A further investment in a piece of plywood and the plugs can be mounted to form an attractive display. The way prices of auto memorabilia are today collecting spark plugs must be one of the few reasonably priced things left to collect. □

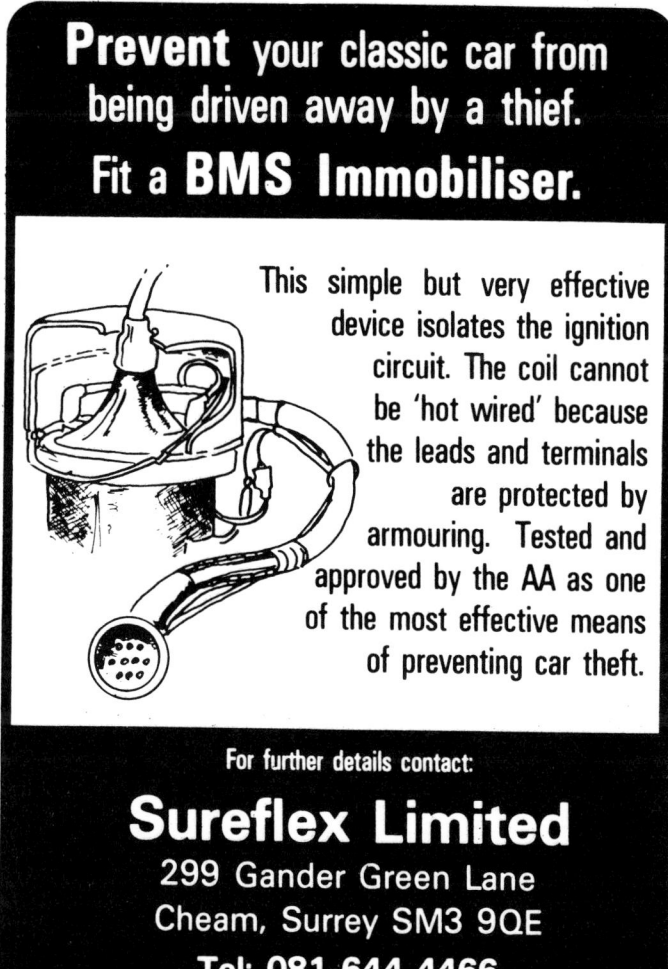

Accumulating Problems

Peter Wallage gives some advice on choosing and looking after your battery

It sits under the bonnet, or in the boot or maybe under the seats and, if it's feeling tired, the car won't start. But apart from remembering to top up the electrolyte every so seldom, most of us don't give the poor old battery another thought. When it eventually gives up the ghost, and won't accept a charge, we get another one from the local accessory shop and, provided the terminals are the right sort, and it fits the battery carrier, price is the main criterion. However, a battery deserves more care and attention than that, both when we're choosing it and after we've bought it.

Lead-acid car batteries haven't changed a lot in basic principle since before the war. They still have a series of negative and positive plates in each cell but battery technology has advanced a lot and, nowadays, batteries of the same power are smaller and lighter — and, with reasonable care, they last much longer, which is why they get neglected more. Batteries used to be rated in ampere-hours, the time it took for the battery to go from fully charged to discharged multiplied by the discharge current, and this is what you'll find quoted in the specification for your classic.

Unfortunately, the new system, rating by reserve capacity and the time taken for the voltage to drop to a certain level under heavy discharge, is not directly related to the old ampere-hour system. You have to go on the word of the man in the battery shop that it's the right one for your car. He'll have a book listing most modern cars but the chances of your classic being listed is fairly slim. Either you take the salesman's word or you pick a modern car with an engine about the same size as yours — not too difficult if it's an Austin-Rover — or you get the salesman to ring up the battery supplier and ask for an equivalent.

Very few batteries get fully charged when they're in use on the car. A survey carried out some years ago by a battery manufacturer showed that, on average, batteries are seldom charged to more than 70 per cent of their rated capacity. When the car manufacturer specified the capacity of the battery for your classic he used volt-amp graphs supplied by the battery maker. These show up another peculiarity of lead-acid batteries. A fully charged 12-volt battery, asked to deliver a current of 20 amps, might well give this at 12 volts. Ask it to deliver 80 amps and the voltage could well drop to around 8.5 to 9 volts, depending on the capacity.

For reliable starting in winter, the battery should be able to crank the engine at 100rpm, so the car manufacturer consulted the volt-amp graphs and the temperature graphs and picked a battery which would crank the engine at about 100rpm in really freezing conditions when it was only 70 per cent charged.

You might think that it's a good idea to fit a much larger capacity battery than the original, particularly if you've fitted more electrical equipment such as a heated rear window which takes a lot of amps, so that you've got something in reserve. This could be a good idea but remember that the

Batteries used to be rated in ampere hours (Ah). This one has a transluscent case so you can check the level of the electrolyte without taking off the caps.

Modern batteries are rated in discharge current and reserve capacity.

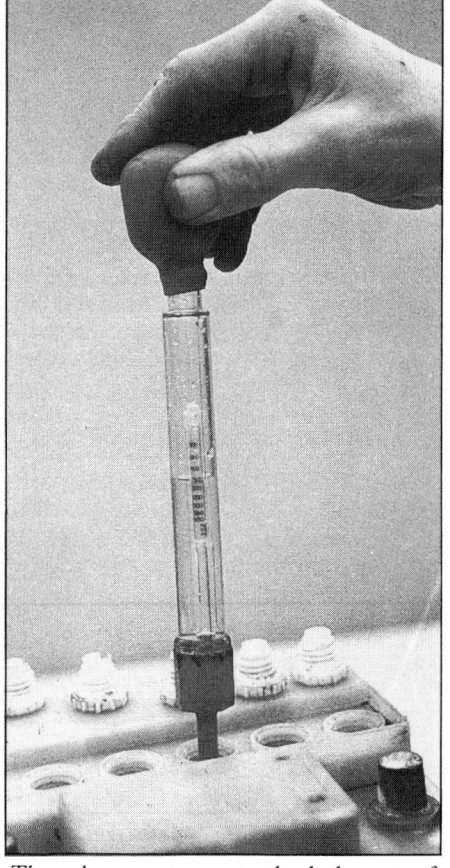

The only accurate way to check the state of charge of a battery is with an hydrometer.

dynamo or alternator on your classic was chosen with the original battery capacity and the demands of the electrical system in mind. You might find that you need a larger output dynamo or alternator to avoid the larger battery always being only about 50 per cent charged. You might not notice this if your car is a first-time starter in summer but a sign that the charging isn't keeping up with the demands on the battery is that the flashers slow down when you've got other heavy-current electrics on. If the battery never gets fully charged the plates harden and the life is shortened.

What about 'cheap' batteries compared with those bearing a well-known name? You

can find own-brand batteries in accessory shops that look very similar to batteries with a nationally known name but quite often at nearly half the price. These are probably quite reputable batteries if they are made in this country or Western Europe. They are in much the same category as own-brand goods on the shelves of supermarkets in that they are made under large-order contracts for battery distributors by the major manufacturers but without the manufacturer's name on.

They are probably not the latest top-of-the-range designs but are perfectly good batteries. The only thing about them is that they are sold to the distributors at a low

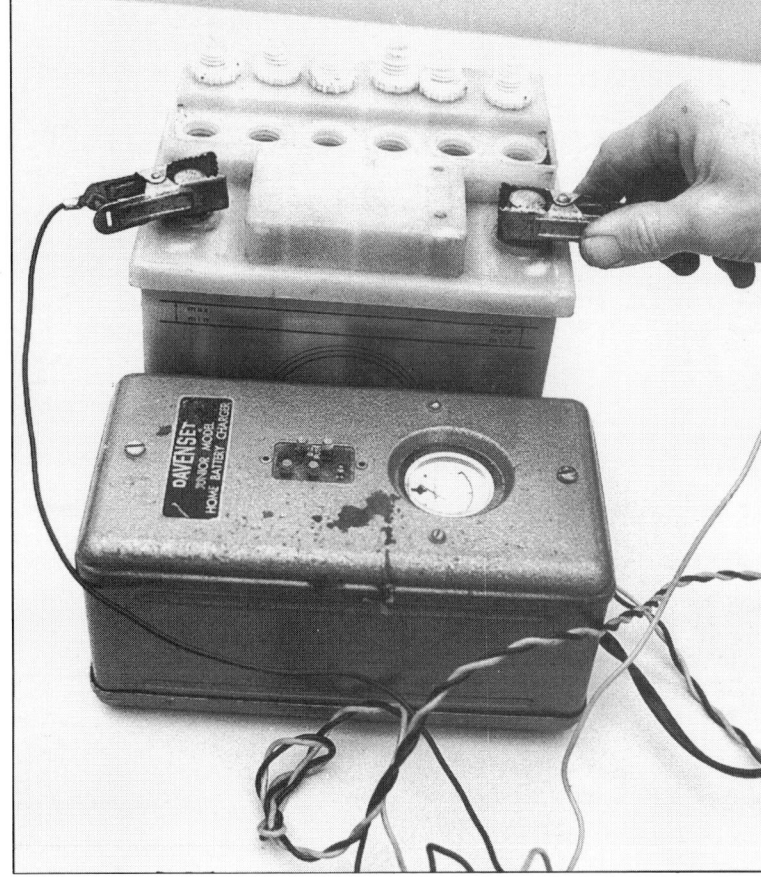

Above; So-called maintenance-free batteries have a separate plug so you can test the specific gravity of the electrolyte. It says "Use no fluids" but, if the level is low, you have to top up but possibly only once a year or so.

Left: Always switch off the charger before making or breaking the connections at the charging clips.

up fully if it has run right down. Battery manufacturers recommend a current for normal charging and this is sometimes printed on the battery label. If in doubt, choose a home charger with a current rating of about 2 to 3 amps, or set a higher output charger to 3 amps and you won't go far wrong. Choose a charger which tapers off the current as the battery becomes more fully charged; then you can leave it on overnight and not worry about it. In an emergency you can boost charge, or fast charge as it's often called, up to about 70 per cent of fully charged at a might higher current but, though doing this once or twice is unlikely to shorten the battery's life, regular use of a boost charger will.

Boost charging needs more care than normal low-current charging. The charger must be one which 'reads' the back voltage of the battery and allows the current to taper off. Boost charging heats up a battery and the temperature should not be allowed to rise about about 43 degrees C (110 degrees F). The better boost chargers have a thermostat probe which can be placed in one of the middle cells and which cuts down the current as the temperature rises. Some batteries have topping-up devices where its impossible to put a thermostat probe in a cell so a careful watch has to be kept on the temperature. In any case, boost charging ought not to be carried out for more than an hour and should be stopped if the battery starts gassing vigorously.

All batteries give off gas when they're being charged, hydrogen from the negative plate and oxygen from the positive plate. These two gases make a highly flammable mixture, potentially explosive if they're in a confined space. You must never charge a battery anywhere near a naked flame or something like an electric fire that glows red and you must always take care to avoid a spark at the terminals. Some nasty accidents have occurred when people either made or broke the connections of the charging clips while they were still live. It needs only a small spark to set fire to the gas mixture and it has been known for the flame to blow back inside the cells and literally blow the battery apart. Always make the connections at the battery before you switch on the charger at the mains and always remember to switch the charger off before you take the clips off the terminals.

Normally you should take the vent plugs or vent cover off the battery before you charge it but there used to be exceptions with some designs of battery which had devices in them to help topping up. Usually these devices used some sort of vent valve at the top of each cell which closed when you took the cover off, or some plastic balls attached to the vent cover so that, when you lifted the cover and laid it to one side, the balls sealed the top of the cells. It's quite a few years now since these went out of production so there probably aren't any, if any, still in service. But just in case, if you've got an old battery that's still soldiering on, make sure that the tops of the cells are free to gas when you take off the cover.

In some accessory shops you see bottles of 'topping-up liquid' which should be distilled or de-ionised water but might be very dilute sulphuric acid. In normal circumstances,

price without guarantee and, should you want to claim under guarantee, the manufacturer won't want to know. You have to claim from the shop where you bought it.

Even starting with a fully charged battery, many classics don't cover sufficient mileage to keep it charged and this is aggravated by the long periods of standing. When a battery is left standing, particularly if it is partly discharged, the plates start to collect a coating of lead sulphate which steadily reduces the battery's capacity and makes it more difficult to charge. Batteries last longest if they are constantly charged and discharged so, if you have an everyday car as well as your

classic, it pays to swap over the batteries from time to time. If you lay the classic up for any length of time, take the battery off and use it turn and turn about with that on your everyday car if it is suitable, keeping the battery which is off the car topped up by trickle charging. Alternatively, if this is not practicable, discharge the idle battery by connecting something like a headlamp bulb across it and trickle charge it up again at least once a fortnight, preferably once a week.

Trickle charging at only one or two amps is undoubtedly the best way to charge a battery, though it takes a long time to charge it

Accumulating problems

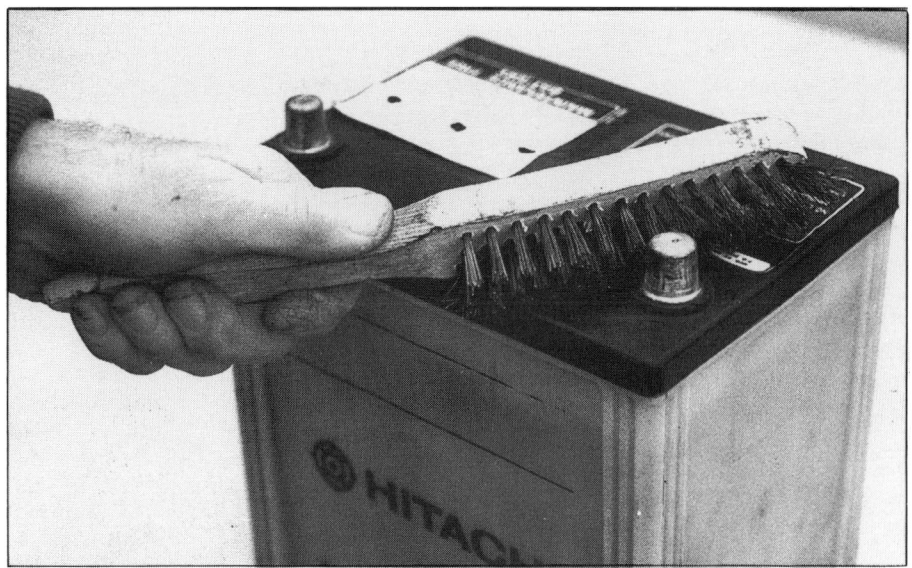

A wire brush is better than emery paper for cleaning battery posts as it is less likely to make the tapers smaller.

you never need to add acid to a battery. The level of the electrolyte drops because of evaporation but only the water content evaporates, the original acid staying behind. The only time you should ever need to add acid to a battery is if you've been clumsy enough to knock it over and have lost some of the electrolyte. In that case you should top up the cells with ready-prepared dilute sulphuric acid of the right strength which you can get from most battery shops. For all other topping-up use distilled or the much cheaper de-ionised water. Ordinary tap water contains all sorts of chemicals which can shorten the life of your battery and even boiling doesn't get them out. I have seen it advocated that you can melt some of the frost from inside a fridge or a freezer but this is a poor second best. When you top up, use a plastic or glass container, not a metal one, or you'll be likely to put metal salts into you battery which won't do it any good at all. You should top up until the level of the electrolyte is just above the top of the plates. Some batteries have transluscent cases so that you can see if the level of the electrolyte is high enough. It helps sometimes to hold a light behind the case.

One way of checking the state of charge of your battery is by taking a reading across the terminals with a voltmeter, which is what dashboard-mounted battery condition indicators are, but a far better and more accurate method is to use an hydrometer to check the specific gravity of the electrolyte in each cell. Get one that has figures on the float, not just plain coloured bands. With a fully charged battery the hydrometer should read between 1.27 and 1.29. A reading between 1.23 and 1.25 indicates about 70 per cent charged and if it is as low as 1.11 to 1.13 the battery is flat. Strictly speaking, hydrometer readings should be taken at 15 degrees C (60 degrees F), so they are not an accurate guide if the battery is still warm after a boost charge or freezing on a winter's morning. If you want to be really accurate, use the temperature conversion table which should come with the hydrometer.

One thing an hydrometer reading on each cell will tell you without fail is if one cell, or a pair of adjacent cells, is breaking down. This is caused by sediment which falls from the plates during normal use, collects at the bottom and shorts out the plates. It usually means that the battery's life is over. You might be lucky and move the sediment by draining out the electrolyte, flushing the battery with distilled or de-ionised water and refilling with acid of the correct specific gravity. This might give you a few months more life but, if it doesn't work, the battery is scrap. Shorting by sediment used to be more of a problem than it is nowadays because many modern batteries have what are usually called envelope separators round the plates which contain the sediment so that it is much less likely to short them out.

Some batteries are labelled maintenance-

This SMMT type of battery lug is far better than the cheaper helmet type.

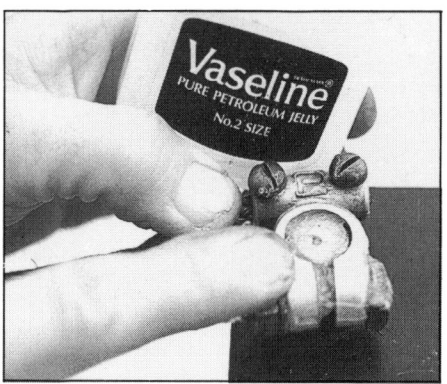

To stop corrosion, keep the lugs and posts clean and tight and smear them with Vaseline.

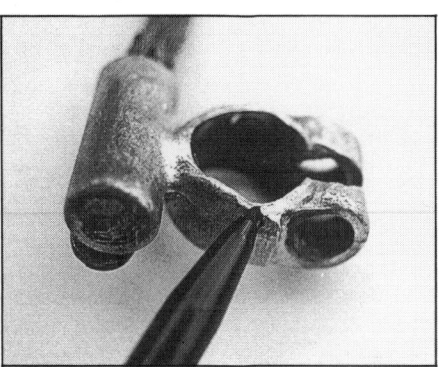

Even with the SMMT type of lug you have to watch for corrosion. If left, it will eat away the lug as here.

free and don't need topping up regularly. Most haven't got filler plugs or a vent cover in the normal way but, if you look closely, you'll find small vent plugs flush with the surface and a plug for you to test the specific gravity, usually with a screwdriver slot to get it undone. Maintenance-free doesn't necessarily mean that you never have to top up the cells. The electrolyte evaporates more slowly than with a normal type of battery but it drops in time so you should check at least once every six months or so and remove the screwed plug to top up if necessary. If any battery needs topping up frequently, suspect your car's charging rate which may be too high. The only types of battery which don't have vent plugs and never need topping up are the non-lead-acid types, usually nickel-iron or nickel-cadmium. They had a

vogue some years ago but seem to have dropped from favour.

Even a fully charged battery won't crank the engine well if the terminals are not clean and tight. Many classics have the helmet type of terminal connector which pushes on the taper of the battery post and is held by a small self-tapping screw. Over the years, with cleaning, these helmets get out of shape and very often don't fit the battery post properly. Sometimes the taper has been scraped out so much that the helmet bottoms on the post without gripping at the sides. You might see advice in some magazines that this can be 'cured' by squashing the helmet with a pair of Mole grips but this is at best a very temporary cure. It makes the helmet grip on two sides only so that it isn't making proper contact and the places where it touches the post lightly burn and oxidise. If you're dead keen on originality and the helmets are worn or misshapen, replace them. My own view is that helmets were introduced because they were cheaper to mass-produce than the SMMT clamp type of connector which is far superior.

Significantly, many car manuacturers have gone back to the SMMT clamp or have changed to the large spade terminal favoured for years by Ford. Whatever types of terminal you have on your car, keep them clean and coat them with Vaseline to prevent those whitish-green crystals which eat away at them. If you do find this forming, clean it off with hot water. □